Operation Saltwater

Operation Saltwater

Fresh Water Crisis Amid Climate Change

CAL RAY

RESOURCE *Publications* • Eugene, Oregon

OPERATION SALTWATER

Resource Publications
An Imprint of Wipf and Stock Publishers
199 W. 8th Ave., Suite 3
Eugene, OR 97401

www.wipfandstock.com

PAPERBACK ISBN: 978-1-7252-6509-7
HARDCOVER ISBN: 978-1-7252-6510-3
EBOOK ISBN: 978-1-7252-6511-0

Manufactured in the U.S.A.   04/08/20

This is a work of fiction. Names, characters, businesses, places, events, locales, and incidents are either the products of the author's imagination or used in a fictitious manner. Any resemblance to actual persons, living or dead, or actual events is purely coincidental.

To my wife of over fifty years, my two remarkable daughters and their husbands, and my five amazing grandchildren.

My thanks to family and friends who have contributed to this narrative with their wit, ideas, and patience. I'm especially grateful to my wife, Gerry, my best critic and support. Also to Grace Mead-Brill for making available her experienced proof-reading skills.

Chapter 1

As the security guard motors his second-hand scooter towards the Nhlabane Nature Reserve, he's distracted by white-backed vultures circling the sky. These feathered scavengers are common, but he doesn't want to come across whatever dead carcass they're devouring.

Turning South, he continues to his post at the Nhlabane water plant. Realizing he's traveling in the direction of the birds, he finds them gathering around the plant facility. Stopping his motorbike, a short distant from the security fence, he's sickened at what he sees.

Vultures are grabbing at four dead bodies: two men at the plant security gate and two women nearby. Shooting his revolver, he attempts to scare away the creatures, but each time birds fly off, others take their place.

He grabs his mobile from his backpack and reports to his superior. "Got dead bodies here. The night guard and three others. White-backs are ripping them apart."

"I'll call the police. You check the plant."

Carefully stepping around the two dead men, he dashes to the building. As he's checking the security mechanisms, he can hear the South African police sirens in the distance.

In the night hours, the multi-tiered floodlights make the busy Port of Richards Bay, South Africa, as brilliant as a category four soccer stadium. The large seaport's a parking lot for dozens of cargo ships filling the harbor with sounds of noisy lifts and banging containers. Leaning on the rails of a cargo vessel, two seamen smoke cigarillos as they watch huge cranes move ocean containers like checkers.

The taller seaman's a brawny Russian in a dark t-shirt and soiled trousers. His scarred face shows no signs of emotion as he puffs away. Because of his boiler room duties, his arms are black from diesel fuel. Also dirty, his oily hair barely moves in the frequent gusts of wind. He's a battle-proven mercenary who conceals a lethal knife inside his shabby engineer boots.

Beside him stands his Aryan colleague, a shorter man. Under his worn cap, his blond hair's also soiled with dirt from the lower decks. His German face has a crooked nose, perhaps from a fall or a fight. Dark clothes help him blend into the night, and his loose shirt hides a 9mm Beretta with a custom silencer. In his pants pocket is a flip phone. He carries it at all times.

The Russian and German have performed below deck duties during this journey, but that is not their mission. Hiding when docked, they've seen little daylight since leaving Rotterdam a few weeks ago. When the ship anchors at various ports, neither man goes ashore. But now, since reaching South Africa, they linger each night on the main deck in the shadows.

Reaching for another smoke, the German's mobile rings. Shuffling the phone from his pocket to his ear, he hears the single word "Gehen," a German word meaning "go." The phone disconnects and he nods to his companion. They hasten to the ship's emergency storage area to retrieve their packs hidden among life jackets, backup generators, and miles of tangled rope. Thoroughly rechecking the contents, they slip the packs on their backs and hurry down the gangplank.

"Should I log you off the ship?" a Black African guard stammers in broken English. "Surfing the local wildlife?" he jokes as his words whistle through his front teeth.

The shorter seaman replies in English with a strong German accent. "There's no need to add us to your log." He drops a half dozen Euro coins into the young guard's hand. "Where would we find this wild life?"

"Port entrance be the place to start." The guard puts his paperwork aside and gives them a smile.

As the men walk to the port gate, they spot a few women with scant clothing pacing around the entrance. The Russian winces at the sight of overweight Black African girls, older Asians with strange wigs, and a few Coloreds. He moans thinking these are the leftovers. He offers the two Black African girls his arm as his German mate hires a nearby taxi. The Russian climbs in the back seat between the two women. The German gets in the front and instructs the cab driver, "Nhlabane Nature Reserve. Gehen!"

After about an hour's drive, the German directs the cabby to follow signs to the Nhlabane Water Plant. Soon, the plant's security lights glow in the dark. The German stops the driver about twenty-five kilometers from the main gate. He exits the vehicle and waves the cab on towards the plant entrance. Arriving at the plant's main gate, the Russian and two girls exit the cab.

Turning to the driver, the Russian demands, "Park by the sea barrier and wait. We'll be here a short time."

Except for routine maintenance on two diesel saltwater conversion pumps, the water plant's deserted due to a holiday shut down. Only a single guard provides security.

The Russian walks his female companions up to the security gate. A Colored boy of about eighteen meets them. Despite his young age, he's wearing a security uniform fitted with a revolver, radio, and handcuffs. His heavy belt also includes a large ring of keys.

"Would you like to share?" the Russian asks the boy as he hugs one of the two girls.

The guard looks at the two girls, but doesn't respond.

"Just open the gate and you have your choice," the Russian urges.

The guard looks down and doesn't answer.

"Fine. I'll have two," the Russian says. He pulls the girls away from the gate towards the shadows. In minutes, the three are groping each other. The guard cannot stop watching them. That's when the young Colored boy feels a pistol touch his neck.

"Open the gate," says a man with a German accent.

At first the guard's unmoving, but when the barrel pushes further into his neck, he grabs his keys. As soon as the gate's unlocked, the German tosses the guard's radio and weapon to the ground. With the handcuffs, he secures the boy to the fence before delivering a blow to his head. Turning his Berretta towards the trio, the German aims and the weapon puffs out two shots, one into each girl's chest. Standing over the girls, he taps them again, so that death is sure. Meanwhile, the Russian puts his black shirt back on and grabs his pack.

Both men move quickly through the plant yard and use the keys to enter the main building. They've memorized the floor plan and location of the two diesel engines. No one's spotted inside the facility, so they hurry to complete their tasks. Reaching the west side diesel motor, the Russian opens the engine oil cap. He pours four hundred grams of ground diamond polishing rouge into the engine crankcase and recaps. Then he opens the coolant reservoir. From his pack he pours two liters of bleach into the coolant system. The German follows the same procedure on the east side diesel.

3

Fifteen minutes later they return to the front gate. The Russian removes the guard's handcuffs, but picks up his weapon and aims at the boy. When the guard's confirmed dead, his German partner fires two quick rounds and the big Russian falls with a thud. The German places his Berretta in the hands of the guard and ensures the guard's gun is in the dead Russian's hands. He rechecks the gate area, locks the gate, and retrieves the Russian's pack. Not an unsolvable crime, but deceiving enough to keep authorities guessing for a while.

Walking to the taxi, he opens his flip phone. He redials the number from his earlier call and utters one word, "Erfolgreich," meaning "successful." Walking to the sea, he breaks the phone in half and tosses both pieces into the water.

Because of an earlier 9mm discharge, he pulls the dead cab driver out from behind the wheel. Hoisting the driver into the trunk, he closes the lid and gets in the driver's seat. Three hours later, he showers, puts on clean clothes, and waits in line at the King Shaka International Airport with e-tickets to Zurich.

A few days later on New Year's Eve, the streets of Bern, Switzerland, are alive with pedestrians and traffic. A German with a crooked nose sits quietly in a taxicab as it struggles through the crowded streets and turns into an alley. The taxi passenger painstakingly counts out the exact fare. Handing the euros to the driver with a German "Happy New Year," he exits the red Fiat Multipla. He watches the taxi leave the alley and then walks to the street. He crosses at the light and enters a poorly lit path into an obscure side street bar.

Nodding to the bartender he strolls past the bar and turns down a narrow hallway. Near the end of the hall, door markings on the left and right designate the washroom genders. The door at the end states "Private." The German patiently waits for the lock to release and then pushes open the private door. Crossing a small office, he's only short steps from a large private meeting room where colleagues are celebrating the New Year.

"Welcome, Karl," says a bearded man in broken English. The man hands him a tall fluted champagne glass with a bubbly liquid. "Made with German water," the man laughs.

Strolling around the room, Karl greets those he knows and introduces himself to others. Many of those in attendance are industrial workers from various European countries, a few well-educated, and all loyal to the

European Union. They've made a pact to infiltrate seawater desalination plants, the facilities which convert ocean water into fresh water. This group has been working in several locations around the globe, including Ras Al Khari, Saudi Arabia; Milos, Greece; California, USA; and the Thames Water Plant in London. The leader's busy joking with a couple of the newer recruits, so Karl waits for an opportunity to speak to him.

He picks up a plate and walks around a long table of Swiss appetizers. Seeing several of his favorites, he selects samples of the various Swiss cuisine: cheese fondue, roti, bundner nusstorte, saffron risotto with luganighe sausage, and assorted cheeses and chocolates. Reaching across the banquet table to add a few specialties from the berner platte, he feels someone grab his shoulder.

"Minister Finn," he responds in Swiss German. "I didn't want to interrupt your conversation."

"I'm glad you're home," Finn says in the same language. "How was your flight back from Durban? I hope you've caught up on your rest."

"The flight was relaxing. Everything you arranged was waiting for me at King Shaka. I'm ready for my next assignment."

"I must say," Finn replies as he reaches for zopf bread and cheese. "You did a splendid job at the Nhlabane plant. According to today's news, the South African police can't find any motive as to the multiple deaths. And they don't believe anyone entered the facility."

Finn motions Karl to a more private area of the room. He whispers, "Even now, after the diesel engines quit running, no one has any suspicion. The South African government will probably blame the diesel manufacturer for faulty equipment."

With a big smile, he pats Karl on the back. "Karl, you've made me and the Eau Suprême members very proud. I hope the next assignment goes as well."

"So do I, Finn. So do I."

At midnight, the group toasts each other and Swiss Minister Finn Schweitzer. Finn motions the group to take their seats.

Speaking in broken English, he commends all the men. "Thank you for your kindness and dedication. As the Swiss Minister of Environment, Transport, Energy, and Communications, I'm very pleased to serve the government of this great country. It's the world's only true democratic nation." He pauses for a minute. "But as an Eau Suprême servant I'm extremely proud to work with every one of you. We have a history together and I'm delighted you've agreed to work with me again. Together we'll make Europe a haven for our families despite the global predictions of chaos, the coming days of climate change." The room explodes with enthusiastic applause.

Seeing a man stand Finn says, "It's late, but I'll take time for your question."

"Minister," the man questions. "Is the plant disruption that occurred in that South African desalination facility our work? So far it appears to be just an equipment failure."

"And that's how it should appear!" Minister Finn grins. "Anything else before we dismiss?"

"Is Amato on board with these disruption plans?" another man asks.

Finn takes a short pause and cautiously words his reply. "Alessia Amato represents the diplomatic European leadership in which everyone wants to believe. We, however, and without her knowledge, are getting the job done." More enthusiastic applause.

An Italian man speaks. "Minister Schweitzer, let me say 100 percent of the bottling companies in Italy are behind your leadership. But I don't believe we've convinced all the bottled water enterprises across the EU."

"Stefano, you're correct. We're continuing to meet with other company leaders. Let's hope the European Climate Change Meeting in mid-March will help show them what is at risk.

"Now, thank you for your attendance. See you at the March meeting. Happy New Year."

As Minister Finn, Karl, and others applaud and celebrate in Bern, a group of executives from bottled water entities are celebrating near the EU headquarters in Brussels. After midnight chimes, Diego Gonzalez, President of the Eurogroup, addresses the hand-picked audience of top ranking corporate leaders.

"Our task is straightforward and simple. In the next ten years, we must protect our countries' fresh water supplies. The exact timing and details of a rising ocean seem to baffle our scientific community, but a rising ocean is certain. In time, fresh water will become the most sought-after commodity around the globe. We must shield our natural water reserves from being depleted or seized by thirsty immigrants desperate to survive. We've already suffered too much from the effects of unwanted immigration and terrorists.

"Therefore, we must expand our bottled water distribution by making our water available to the Middle East and African community. Access to fresh water will significantly postpone unwanted immigration. Providing controlled water delivery will also give the EU an influential position in the policies of MEA countries. And who knows?" he smiles shrewdly, "It might also make us very rich."

As the room applauds, President Gonzalez acknowledges a raised hand.

"Mr. President, I believe we have a business challenge before us. The desalination process is already crippling the bottled water business in certain international regions. Won't it be a bigger problem as the oceans rise?"

"Thank you for that question," President Diego responds. "The desalination process is still technologically too expensive for many localities. We expect future advances in new technology will make it more affordable. The structure of the Eau Suprême organization, under Alessia Amato's direction, is to protect Europe's fresh water resources and related businesses.

"The challenge I see to continuing Europe's profitable business is the ability of our water industry to provide water in larger quantities. Drinking water is basic, and we do an excellent job of providing bottled water worldwide. Our arid neighbors need water for more than drinking. We'll discuss these and other climate change issues at our March meeting. I hope to see representatives from every bottled water enterprise.

"Thanks for your participation, and Happy New Year to Europe."

Chapter 2

BRADLEY MOVES HIS STEP ladder closer to the Christmas tree. The nine-foot white needled tree stands elegantly in the thirteen-foot-high foyer of his Springfield, Ohio home. Bradley, a high school senior a little over six feet tall, might not have needed the ladder, but he's extra cautious removing the decorations. His mother, Rosemary, purchased the huge imitation tree a few weeks before her acute leukemia diagnosis. She loved the tree so much, that Conley, Bradley's father, agreed to leave it standing as she fought the disease at home. Unfortunately, because of her chemotherapy, bone marrow complications took her life. Bradley's proud that his dad still assembles her special tree each Christmas.

Since his birth in Columbus, Bradley Truman has grown up in a loving and faith stimulating environment. His father was a senior attorney in a prestigious Columbus firm, so his mom was a stay-at-home Mom devoted to their only child. Bradley seldom suffered disappointments in life until his mother died. He stayed at her bedside during the last few days of her life and made several promises to her. Promises he still intends to keep.

As he finishes disassembling the artificial tree limbs for storage, Conley descends the short stairway from the upper level of their split-level home. Conley's the same height as Bradley, but his graying hair and facial lines resemble a grandfather more than a father. He resigned his successful partnership in Columbus when Rosemary became ill. The family moved to Springfield to aid in Rosemary's medical care. Launching out on his own and working from home, Conley's been a successful breadwinner and companion. He and Bradley have become close.

"Brad, you've done a great job. Christmas break's almost over. Are you yearning to get back to school?" Conley teases.

"Dad, you know I miss science class. My project's pieces are working together, but I have a lot of testing to do. There's a chance I can win an award in the Ohio State Science Fair competition."

"Brad, I just talked with your principle. Let's sit for a minute. I've got some disappointing news."

"You were talking with Mr. Maxwell?"

Conley smiles. "Let's just say that occasionally everyone needs an attorney." His face turns solemn. "Mr. Galloway won't be returning after the break. He's taken another teaching position at a Dayton community college."

Conley spots the distress in Bradley's face. He says nothing, but can sense his son's brain going into overdrive. Conley waits. He knows Bradley's counting on Mr. Galloway's support for the state competition.

"That means I'll get a new teacher when break's over," Bradley laments. "A new teacher may have no interest in my water project. Without a sponsor, I might not qualify to enter the competition." He pauses for a minute. "Dad, this is a five-star disaster."

"So what do we do in such an event?" Conley says tongue-in-cheek.

"Well, first we pray," Bradley says, trying to keep from showing a grin, "and then we eat a Jim Dandy sundae with lots of nuts. On a day of such tragic news, I think I should get to pilot the LTD." Bradley lets a big smile sneak out. "It's in God's hands. What else can we do?"

"Keys are in the kitchen. Meet you in a second."

Bradley's delighted to drive the vintage Ford LTD. With a junior license he's been driving their 4-door F-150 pickup. It has four-wheel drive and plenty of power, but doesn't have the pizazz of the robin's egg blue 1971 classic with a convertible top. They don't drive it much in the winter, but Bradley expects this special circumstance to put him behind the wheel. It'll be his job to wash it when they return, but he's sure it's worth it.

Conley grabs his winter coat and soon slides into the front passenger seat. Bradley adjusts the bench seat, checks the mirrors, and starts the engine.

"Let's pray first," Conley encourages as the engine hums. The two bow their heads and thank God for Mr. Galloway's unexpected exit. Conley asks for divine direction regarding Bradley's science project. Bradley especially prays for the support of his new science teacher, whoever that may be.

"Got your license with you?" Dad asks. Bradley nods.

"Then let's take her out captain," Conley exclaims, and Bradley backs out into the street. Arriving at the restaurant, they're soon eating two large ice cream sundaes. As they enjoy the dessert, they discuss all the "what ifs" this change in teachers may mean.

Arriving early at school on January 2nd, Bradley hurries to the top floor science room. His water project parts and pieces are laying all over the lab. He's worried his paraphernalia isn't going to make a good impression. Busy packing up his equipment, an unfamiliar female enters the room.

"Hi. I'm Brad," he says, only half paying attention. As the young-looking girl wanders around the room, he continues working and talking. "I know most of the high school students, but I don't think we've met. Are you new to Springfield High?" Her continued silence causes him to put down the pieces in his hands, and he turns towards her.

Instantly there's a lump in his throat. He's charmed by her beautiful brown eyes, her adorable unblemished fair complexion, and her short dark hair. Wearing black slacks and a frilly white blouse with a burgundy jacket, he thinks she's looking very attractive. Her face looks young enough to be a sophomore. As she smiles at him, his eyes fix on her glossy lips. He knows his staring is beginning to look stupid. He wants to say something, but he can't get his mouth to connect with his brain.

At last, she speaks, "Do you enjoy science?" As their eyes meet, he struggles to answer.

"Oh, yes, I think. I mean, yes, a lot. I love science. I think science is fantastic," he blurts, his face turning redder as he stammers.

"Is that why you're here so early?" she questions.

"Well, a new science teacher starts today. My guess is a retired teacher covering for Mr. Galloway's surprise resignation. Anyway, I don't want to bug a new teacher by leaving my project all over the lab, so I'm cleaning up my mess," Bradley laughs as he grabs a piece of pipe to show her.

"What's your science project?" she asks, keeping a serious tone.

"I'm demonstrating the lengthening of water molecules," he explains. "It's not a new concept, but I'm hoping to stream water uphill without pressure. My goal is for water to run uphill at least 12 centimeters."

When she doesn't comment, Bradley tries to be hospitable. "I'm guessing you must be new. I'd be happy to show you around." Then without thinking he adds, "It would be my honor to escort you anywhere." Bradley lowers his chin. "I'm sorry. I'm sounding like a chump."

She tries not to laugh at his foolishness and says, "Beth. You can call me Beth. You're very sweet. Perhaps you can escort me sometime." She smiles at his embarrassed facial appearance and checks the clock. "Are you a senior?"

"Yes. I graduate in June."

"Do you have plans for college?"

"My dad wants me to go to law school, but I'm determined to stick with science. I'll probably go to Dayton for undergraduate studies. I'm not sure which science I want for a major, so I've made no plans beyond that."

No stranger to schoolboys, she's accustomed to male swaggering and arrogance. So many men have pestered her for attention, her attitude towards the male gender is apathetic. However, there's something about Bradley's unassuming manner and boyish innocence that attracts her. She's taken by his medium length curly blond hair, handsomely sculptured face, and tall athletic build.

"What's so important about this water project?"

"Mr. Galloway was sponsoring me in the Ohio State Science Fair competition. I hoped for an award, but now I don't know. Depends on whether the new science teacher has any interest in my project." Bradley returns to storing his equipment.

The first period alarm bell sounds. As Bradley picks up his books, Beth says, "From now on, please call me Ms. Williams. I'm Mr. Galloway's replacement." She gives Bradley a wink and his face turns red again. "You're a very handsome young man, even with a blushing face. And I meant what I said, thanks for being so sweet to me this morning."

When the third period begins, a portion of the senior class enters Ms. Williams' science room. Bradley brings a shiny red apple and leaves it on Ms. Williams' desk. Strolling to the back seats, he ducks behind the other students.

"Good morning class," Ms. Williams begins. "I'm taking the place of Mr. Galloway. Mr. Galloway has accepted another position and won't return this year. My name's Elizabeth Williams. I don't think I need to write it on the board for seniors. While on this property or during any school activity please call me, Ms. Williams. But, if we meet at a grocery store or in a restaurant, you're welcome to call me Beth. Are there questions?"

A student named Martha raises her hand. "When did Mr. Galloway resign? He never mentioned this before we went on break."

"Mr. Maxwell informed me Mr. Galloway accepted a new position after break began. No one knew about his resignation in advance." Ms. Williams waits for other questions.

A boy in the middle section raises his hand. "Are you single?" The other boys, except Bradley, laugh and the girls respond with groaning noises of disgust.

Another senior boy makes a remark, "She must be single. She's not old enough to go on a date."

"Had enough fun?" Ms. Williams' tone becomes serious. "I started my college education at Robert's Wesleyan in New York State at age sixteen. At eighteen, I transferred to Cornell. After I graduated from Cornell, I enrolled at Stanford and completed my master's degree in science. I left Stanford at age twenty-two and taught in Pittsburgh for two years." With a smile she adds, "Yes, I'm single, but I don't date anyone under age eighteen or any high school students." The whole class laughs.

"Now, all this week I'll be giving you pop quizzes on the material Mr. Galloway should have covered so far this year. I understand he often drew outside the lines, so I want to be sure we don't miss information you should have already discussed. I'll grade the pop quizzes, but I won't include those grades in the marking period scores. The quizzes will simply help us find our place in the overall curriculum. If you do poorly, these tests may help you understand what areas you need to review.

"Can anyone tell me the last chapter Mr. Galloway reviewed in your textbook?" After students suggest various chapters, a consensus develops and they decide where to begin. For the next several minutes, Ms. Williams asks questions of various students to see what they've learned. The exercise helps her remember the student's names.

Two minutes before the lunch bell she states, "It's my understanding all of you must present a science project at the school art and science exhibition next month. It determines a major portion of your science grade." Heads nod around the room.

"Is there anyone who's planning to place their exhibit in the state science fair?" Six students raise hands, including Bradley.

"I'd like to meet with you here during lunch. I'll bring my lunch and, if you bring yours, we can review your projects. If you prefer, you can grab a bite at the cafeteria and then meet with me. Any day this week's fine." The bell rings and everyone disperses except Bradley.

After the classroom empties he ambles up to Ms. Williams' desk. Without looking directly at her, he says apologetically, "Ms. Williams, I'm so sorry for that stupid question Terry asked. He's always trying to joke around."

Ms. Williams picks up the apple and studies him with a concerned face. "Bradley, I need you at your best. Not apologizing for other students, not hiding in the back, and not pretending you need an excuse to talk to me. I need to see an industrious senior doing his best to excel in science. I'm already your friend and your teacher. Do you understand?"

"Got it," Bradley responds, raising his head. A big grin comes over his face. "I brought my lunch today. I'll be back soon." As he walks out of the room, Linda Evans, an attractive and popular senior girl, comes in with her lunch bag and a soda.

"Let's see," Ms. Williams speaks cautiously. "You must be . . . Karen. No, you sat over by the window. You're Linda, Linda Evans."

Linda nods and smiles. "How'd you get through all that college so fast?"

"Well, let's keep this between ourselves." Ms. Williams says as she sits across from Linda. "I have a photographic mind. If I read something once or see a visual image, I never forget it. It doesn't work so well with other senses, for example hearing, but everything I see is mine forever."

"Wow, that's incredible," Linda remarks as Bradley comes back into the room.

"Bradley, come sit with Linda and me. We'll review her science project first." Ms. Williams motions to a chair near the two of them.

Linda opens her backpack and pulls out a handful of small diagrams. She hands them to Ms. Williams and says, "I plan to set up two display boards, one depicting direct current and the other alternating current. My dad's an electrician and he'll help me set up the right switches and voltage meters. I'll also have a photo album by each display. The photos will picture common uses of the different currents so everyone can easily understand."

"Linda, that sounds good. I'm glad you're working with your dad. However, I'm not so sure that's a strong enough demonstration for the state science fair. I expect the state fair to be very competitive and favoring leading-edge science exhibits."

She turns to Bradley and asks, "Any ideas?"

"The last three years someone has prepared an electric current exhibit for their senior science project," he responds. "I like Linda's idea about using a photo album, but in the two state science competitions I've attended, no one has entered an exhibit displaying older technology."

Bradley turns his head towards Linda. "Perhaps if you compared it to wireless devices like Bluetooth, you could show how electrical connections have changed in the last years."

Linda thinks for a minute. "I don't think my dad can help me with that," she says with concern.

"Yeah, but how about somebody from Mr. Smith's computer class," he replies. "They must discuss various wireless technologies." Bradley takes a bite out of his sandwich and waits for Linda to respond.

"I would like to enter the state competition. I'll talk with Mr. Smith and see what he recommends." Linda gives Bradley a big smile. "Thanks," she adds, putting her hand on his shoulder.

Ms. Williams looks at Linda and says, "Get back with me on what you decide. It's completely up to you." Linda eats without responding.

"So Bradley, let's talk about your state exhibit," Ms. Williams says with enthusiasm. "And Linda, please make any helpful comments."

"Well, my basic goal's demonstrating how water stretches. I want to show water climbing over an obstacle. Everyone knows you can siphon water with a hose to move it from one level to another. I want to move water uphill without a hose and without pressure."

Linda listens carefully and asks, "How high do you want to move it?"

"My goal's twelve centimeters."

"How high have you gotten?" Linda replies.

"Well, before break I put my complete exhibit together in this lab. Based on my first test, it was a little over two centimeters. I need more testing. I'm studying microwaves to more precisely control the source of heat."

"So you've been to the state exhibitions," Ms. Williams asserts. "What makes you think your display will be competitive?"

"Mr. Galloway asked me the same thing. I'm not sure. I'm hoping the manipulation of water molecules will be enough to interest the judges." Bradley tries not to look into Beth's eyes, so he deliberately looks away when he speaks to her. He's afraid he'll embarrass himself in front of Linda.

"How will you display it?" Linda asks. "My biggest concern is that people will not even open my photo albums."

Looking at Linda, Ms. Williams inquires, "What do you recommend?"

"Some kind of artwork display, or maybe a few gadgets so people can understand what the water's doing. I don't know," she shrugs.

Standing to drop her lunch bag in the wastebasket, Ms. Williams turns to Bradley. "Do you have any gauges to depict water flow, molecule changes, or anything like that? If not, the right diagrams might aid your exhibit."

Bradley thinks for a moment and looks at Linda. "Thanks, I like that idea. The better state exhibitions often have elaborate graphics. I've been so busy working with the water I haven't given displays any thought."

"Okay, I think we've made progress," Ms. Williams says. "We still have about twenty minutes before the next class, anything else we want to discuss?"

"Ms. Williams, sometime I'd like to discuss something in private. Girl talk," Linda says.

Bradley stands. "Hey, I've got to run. Thanks to you both." Bradley pitches what's left of his lunch into the wastebasket and disappears out the room.

"What is it, Linda?"

Looking around, Linda makes sure they're alone. "It's Brad," she whispers. "I need a woman's opinion." Ms. Williams moves her chair a little closer to Linda.

"I've wanted Brad to ask me out. I've dropped my share of hints, but I'm invisible to him. Today's the first time we've ever been . . . alone . . . together . . . whatever."

"Who does he date?"

"No one," Linda shrugs. "Half the girls in the senior class would date him, but he's not dated, gone to a game, or dance . . . or anything. He's handsome, friendly, comes from a nice home, and he's everybody's friend. Honestly, he's one of the nicest guys in our school."

"So what's holding him back?" Ms. Williams' inquires.

"I don't know. His mom died about five years ago and he seems over it. He goes to church, but I don't think his church prevents dating. He's mature for his age. In fact, he's more mature than most of the guys in our class. Either he doesn't know how to be around girls, or he's . . . You know . . ."

"I understand Brad gets top grades," Ms. Williams replies. "Maybe he doesn't want to get sidetracked. I did very little dating in college. Perhaps after graduation you'll see a change."

"Maybe. I hope so." The bell rings and Linda hurries off to her next class.

On Friday, Ms. Williams reviews the state exhibits again with the senior class. "Let me go over the state exhibits one more time. For the Ohio state exhibition, I've got Linda, presenting an electrical exhibit; Terry, presenting a blood evaluation exhibit; Martha, presenting new technology emergency care methods; Karen, displaying deviations of the flu virus; and Bradley, exhibiting the stretching of water molecules. Are there any other exhibits planned for the state science fair?" No one responds.

"Okay, then. Let's set up those state exhibits in this room on Monday." The bell rings and everyone leaves except Bradley.

"How can I help you, Mr. Truman?" Ms. Williams gives him a big smile.

"You remember the first day we met?" Bradley says, enjoying her brown eyes and curly lashes. "I was putting away the pieces of my exhibit. It's all here in the back closet, but when it's set up, it takes up a big chunk of this room. I think I need a different place."

"Where was it before you brought it to school?"

"While it was warm, I had it in our garage. We have a three-car garage, but only two vehicles. This time of year it's too cold."

"Any other place at your home?"

"I'll talk with Dad. We can heat our lower level, the basement. Maybe I could use it, but a lot of my mom's hospice equipment's stored there. I'll let you know on Monday."

"Bradley," Ms. Williams says, "I've got an appointment with Mr. Galloway this weekend. I'm hoping I'll get the inside story on your project when I talk to him." Her eyes twinkle as she speaks.

That evening over dinner, Bradley asks, "Dad, what's the chance I can set up my science exhibit at home again?"

"Well, the garage is definitely too cold. Your water display'll freeze. Why aren't you leaving it at school?"

"There isn't enough space in the science room for my water exhibit and the other student's projects."

Conley takes another bite while thinking. "Well, the only other place is the basement. All the medical equipment we bought for your mom is still down there. There's a hospital bed, heart monitor, heart defibrillator, plus a bunch of other stuff. If you can find someone who'll use that old equipment and will pick it up, the basement's yours until you graduate. Then it's mine again. Okay?" The two high-five their hands and the deal's done.

"One more thing, Dad." Bradley looks his dad in the eyes. "I'll doubtless get a few visits here from my new science teacher, Ms. Williams. She'll want to review my project."

"No problem. I'd like to meet Mr. Galloway's replacement. Are you getting along?"

"I think so. I want you to meet her."

"Well then, let's plan to invite her for dinner. We'll do that right after you get your exhibit ready."

Two weeks later, as Conley's helping his neighbor, Jaqueline, find her way around their kitchen, Bradley swings by the refrigerator for a coke.

"What's that ungodly sweet smell?" Conley exclaims laughing.

Bradley stops. Conley and Jaqueline are laughing so hard tears are streaming down Jaqueline's face.

"Bradley, my dear," Jaqueline says, trying to calm herself. "You're wearing enough cologne to drown a horse. A woman likes a hint of perfume."

"I suggest you wash your face and change your shirt," Conley chuckles. Bradley runs red-faced back upstairs to the echo of more kitchen laughter.

While Bradley's cleaning up, the Truman's front doorbell rings. Conley opens the door and sees a strange young woman. "We've already made plans for the evening. Perhaps you can bring your sales stuff around another time."

Bradley runs to the door and says, "Dad, let me introduce, Ms. Williams."

Bradley stands in the doorway, admiring her. She's wearing a cute red hat with black mufflers and her winter coat's adorned with a multi-colored scarf. Her coat's unzipped and Bradley can see an attractive embroidered ivory blouse and gray pleated dress slacks.

"You look absolutely splendid," he says holding the door open. She responds with a wide grin. He hopes the sparkle in her amazing eyes is just for him.

After a moment he steps aside. "Please come in Ms. Williams, we've been expecting you."

As she steps into the home, her countenance brightens with the elegance of the foyer. "What a beautiful entrance you have, Mr. Truman. The decor is delightful."

"You're looking at the talent of my late wife, Rosemary. We've made very few changes since she passed." Looking a little humiliated, Conley adds, "Sorry about the misunderstanding at the door, Ms. Williams. I thought you were one of the local teenagers. Please call me Conley." Conley holds out his hand as Bradley hangs up her coat and scarf.

She responds, "Please don't worry. Someday my face will catch up to my age." After the handshake she adds, "And please call me Beth, both of you. We're not at school tonight."

Hearing a female voice, Jaqueline comes out of the kitchen. "Dinner's ready. Should I put it on the table?"

"Yes, please, and I hope you'll join us," Conley responds.

"Beth, meet our cook for the evening, Jaqueline Penn," Conley explains. "Her husband works second shift this week, so I asked her to cook for us. She was a close friend of Rosemary's and tonight she's our chef."

Bradley smiles. "Dad and I manage all right for ourselves, but not so much for guests."

The dinner conversation keeps to pleasantries: the winter weather, Jaqueline's family, a little history of Springfield High, and Beth's first weeks at school.

When Jaqueline picks up the table dishes, Beth stands and says, "Let me help." Conley, however, motions for her to sit back at the table.

"Dad, I need to prepare my exhibit, so I'll leave you two to talk. I should be ready for Beth in about fifteen to twenty." Bradley heads towards the basement.

Conley looks at Beth. "You're not at all what I expected. Please tell me a little about yourself."

"I completed my undergraduate degree at Cornell at age twenty. Then I finished my master's degree at Stanford. The last two years I taught high school science in downtown Pittsburgh and now I'm here."

"I'm impressed. How did you enjoy living in Ithaca? I was at Cornell for two years before I transferred to Harvard Law School."

"Actually, my first two years were at Roberts Wesleyan College in Rochester. I transferred to Cornell in my junior year."

Conley pauses for a minute. "I hope you're not offended by this question. I believe Roberts is a Free Methodist school. Are you Free Methodist?"

"I am, I guess. My dad, before he left, took me to a Free Methodist church. He graduated from Roberts and sent me there to start college. College life gives a person's religious upbringing a good beating," she sighs, "but I'm hoping to get my spiritual life back together once I find a place to settle down."

"Bradley and I attended a Church of God in Columbus while Rosemary was alive. She grew up in that church and wanted to keep her friendships. When her health failed, she became an outpatient at Springfield Regional Cancer Center, so we moved here. After she died, we settled on a good Presbyterian Church a few blocks from here."

"I'm so sorry about your wife."

"Thank you. It's been a challenge, but Bradley and I have become quite close." Conley questions, "You said your dad left? Did he pass away?."

"No. But when I moved to Roberts at sixteen, he disappeared. I haven't heard from him since."

"What about your mom?" Conley says with a concerned look.

"Not much to tell. My mom left home when I was twelve. I've never heard from her either."

Conley changes subjects and starts talking about Bradley. "Beth, I'm afraid my son has a huge crush on you. It must be obvious."

"It is." She blushes. "But, he's been a perfect gentleman. He's a fine young man and will someday make a wife very happy."

Before Conley can comment, Bradley appears in the room. "It's ready." Beth excuses herself and follows Bradley to the home's lower level.

Chapter 3

WALKING INTO THE BASEMENT, Bradley speaks in a quiet tone. "Did Dad give you the third degree?"

"Your dad's great. He just wants to know who's influencing his son."

Gazing at the water exhibit apparatus she says, "Let me see if I understand what's happening here."

She walks to the furthest point and starts identifying Bradley's equipment and process. "This barrel's your water source. It provides a stream of water into this . . . Is this PVC pipe?" she asks. Bradley nods. "Okay. Then the PVC pipe directs the water to . . . It looks like an old rain gutter."

"Right again," he chuckles.

Beth continues. "The water travels down the rain gutter, over this metal . . . device, into more PVC pipe and finally into a collection tank. How'm I doing so far?"

"Perfect!" Bradley exclaims with a thumbs up. "The device is an adjustable conduit, like a draw bridge. When I crank it up, it provides an alternative route for the water. If part of the water flows over the bridge, then I have stretched the water molecules to a measurable length." Bradley points to the attached metric measuring scale.

"So you're calling that device a bridge," Beth says, walking back and forth. "Now where's the magic? What makes the water molecules stretch?" she inquires.

"Look at this part of the water channel," Bradley points to a shiny section in the rain gutter. "This section's coated with high grade nickel." He steps over to where the water exits the PVC pipe and lights a Bunsen burner.

"Now as the water heats it affects the hydrogen and oxygen molecule vibrations dissociating them from each other. As the molecules flow over the nickel catalyst and cool, they recombine into symmetrical or asymmetrical lengths.

"That's why the bridge device can't block the water flow. There must be an alternate route for the shorter asymmetrical molecules. I'm hoping very long symmetrical molecules will form, and I'll be able to measure their length based on the bridge's height."

Beth looks around and grabs an old folding chair for a seat. "I'm very impressed. Sounds like you've given this homemade design some thought, but a few things concern me. Are you ready for a teacher's thoughts?"

Bradley looks into her dazzling eyes and replies, "Lay it on me, teach."

"How will you present all that molecule stretching information? Can you prepare something people can look at, a diagram, power point, or other explanation? How about gauges? We talked about this with Linda."

Bradley opens his mouth to respond.

"Wait a minute," she says firmly. "I'm transmitting, not discussing."

Walking over to the Bunsen burner, she says. "This open flame bothers me. Let's think about how else to heat the water."

Turning back to Bradley, she warns, "There's a lot of good science here, college level thinking. Nevertheless, I'm worried it won't play well in the state competition. This pile of PVC pipe and rain gutters needs some flair, a sexy look. You need something to 'wow' your audience and the judges."

Bradley grins. "That's the nicest thing you've ever said to me." They're laughing at his comment when Conley comes down the steps.

"You're having too much fun down here," he quips. Looking at Beth he says, "Can you make anything out of this collection of water and spare parts?"

"Actually, I think Bradley's premise is college level, but he's got a little work to do. And I should get home," she exclaims.

While Bradley gets her coat, she says, "Mr. Truman, I'll need extra time with Bradley before the high school exhibit. Is it possible I can spend a Saturday here in the next couple of weeks?"

"Well, I guess that'll be okay. Give us your preferred schedule, and I'm sure we can make it work."

Bradley walks Beth to the door. "Thanks so much for coming. I'll brainstorm your ideas and see what I can do. See you Monday."

As Bradley closes the door, his father shows a serious face. "Your teacher's a well-educated and a very attractive woman. I can't believe she looks so young. I hope you're not getting too smitten. The legal implications of a teacher-student relationship are scary."

"Dad, you're too late. I've already asked Beth for a date. Well, not exactly, I sort of told her I wanted to date her."

"What?" Conley exclaims. "What are you thinking?"

"The first day after break she came into the science room. I thought she was a transfer student. She was awesome, and although I got tongue tied, I said I'd like to show her around. That was before I found out she was the new Mr. Galloway."

"Was that embarrassing?"

"I had a red face most of the morning. So far, no one's found out about it."

"Have you talked with her?"

"Yes. She asked me to be a model student and not think about it again, but she didn't say she wouldn't date me after I turn eighteen and graduate. I promised Mom not to date in high school, but I'm thinking ahead," he responds with a silly grin.

Conley smiles and shakes his head. "You picked Ms. Williams to be the first girl you ask out. I guess you only want the best." He puts his arm around Bradley. "Just be careful. If not for yourself, then for her."

The first Saturday in February, Ms. Williams returns mid-morning to the Truman residence. The weather's in the sixties and the garage doors are open. Conley's cleaning and polishing his LTD.

"Hey, Beth," Conley yells. "Bradley's not here. He'll be back any minute." He puts down his polishing rag and says, "Go into the kitchen through the garage."

Beth, followed by Conley, steps into the kitchen and they drop their coats on a chair by the island. "I want to talk with you about Bradley."

"Of course."

"I'm worried he's become infatuated with you. I've been talking to him about the consequences of a teacher-student relationship."

"I can't be his first girlfriend, can I?"

"I think so. I've never worried about him and girls. He and his mom had a heart to heart talk about dating before she died. He's never had a date to my knowledge. In fact, he's never mentioned a girl's name before. That boy has always saved his money and had his head stuck in a science book.

"But now, he talks about you constantly. He tells me what you wear to school, what jokes you tell in class, and how smart you are about everything. My biggest clue, however, was how much of his savings he spent to impress you with his new exhibit framework."

"Did he tell you about the day we met?"

"Yes. How he indirectly asked you out?"

Beth replies after a little pause. "I think you'll understand what I'm about to say," she says seriously. "There's something unique about our friendship, something natural, but also something spiritual. I don't know what it is, but God's destined us for something, at least for a while. Not sure it means romance or marriage, but it means something. We both need your prayers and advice."

"Please, for your sake, and his, be careful," Conley pleads. "I can't help but worry for you both." Hearing the F-150 enter the garage, they end their discussion. Bradley flies into the kitchen with a big tub of Kentucky Fried Chicken, soft drinks, and bottled water.

"I don't know when you'll be hungry, so I picked up lunch. We can eat now or throw it in the refrig until later." Bradley sets everything down on the counter and looks at her. "You seem laid back today. I've never seen you in blue jeans. Looks sorta 'down home'. I like it."

"I want you two to stay out of trouble," Conley says, half joking. "After I'm done with the LTD, I'll be working upstairs on a few legal matters. Oh, and I'll be taking a little lunch to the garage with me." Conley grabs a couple pieces of chicken and puts a large spoonful of coleslaw on one of the KFC paper plates.

"Beth, you want anything?" Bradley asks as his father disappears.

"Not now. Let's get started," she says, stepping towards the basement stairway.

Walking down the basement stairs, Beth's eyes pop. Bradley's whole exhibit has changed. She walks to the water reservoir barrel. It hangs from two welded black steel bars with an adjustable height handle. The black barrel's encircled with bright orange, red, and yellow lightning bolts. The PVC pipe has become black iron pipe; a painted steel channel replaces the rain gutter.

"This is incredible. I love it," Beth exclaims, pointing to the artwork. "Look at the waves painted on the pipes and channel." She walks over and brushes his arm with her hand. "I'm overwhelmed."

"Look at all the features," he brags. "The frame holds all the pieces. It's on wheels and separates into three sections for easy assembly." Bradley points to the locking wheels and pin system attaching the frame together. "Once assembled, like it is now, I can unlock the wheels and move the whole exhibit."

Mounting his laptop on a metal shelf welded to the frame, Bradley starts a computer display of the moving, stretching, and changing of the molecules. The program runs three minutes and then loops again.

"Oh, I really like that," Beth says as the program automatically replays. "I'll try to get you a bigger monitor. Maybe we can use one from the science room. You didn't build this all by yourself. Is this what you bought with your savings?"

"IronWorks Inc., a former client of my dad's, built the frame. A young friend of mine painted the graphics. I worked on the computer program with Karen. I provided her with the basic information on molecule stretching and she did the rest."

"So what's next?" Beth asks. "Are we ready to get this black beauty running? Thought any more about the heating process?"

"I mentioned before I've been researching microwaves. A standard microwave oven uses about two point five gigahertz to heat objects, including water. However, water stretching requires a much faster vibration, somewhere around one hundred terahertz. I've rigged up an industrial microwave so I can adjust the molecular vibration into the terahertz range. We're testing it today."

"I'm ready." Beth grins.

Bradley takes about five minutes to get the microwave clamped on the frame. He plugs in the unit and releases the water. As the water flows down the channel, he plays with the vibration frequency.

"Bradley, how 'bout I lift the bridge a little." Beth suggests. She slowly cranks up the bridge height. She stops at three centimeters.

"Water's flowing over the bridge," she states with enthusiasm. "A little more?"

"Let me see if I can increase the stretching," Bradley responds. As he continues to adjust the microwave controls, most of the water diverts over the bridge. "Okay, take her up."

Beth moves the bridge higher and water continues to climb over six centimeters. Bradley's adjusting the microwave and nods to Beth to increase the height again. Reaching eight centimeters he says, "I don't think I can fine tune this microwave any better."

"Well, that's about three inches," she says. "If I remember, it's more than twice what you could achieve with the Bunsen burner."

"That's right. Not as much as I hoped, but enough to prove my theory. Especially since I have this sexy new frame," he laughs.

"Did you decide on what gauges to add?" she asks looking around.

"Since little or no water pressure helps substantiate my premise, I thought about floating mini-marshmallows down the channel. That will show how slowly the water moves." Beth's mouth drops. He chuckles at her look of incredulity, and she lightly punches him in the arm.

"I built a wire mechanism of two small loops, like a pair of monocles, to check the speed of the water. I'll mount the loops at the end of the nickel plating, attach them to a small battery, and connect a twelve-volt gauge. I'm hoping it'll measure the speed of the stream."

Bradley grabs a small gauge with wiring off a shelf. As he prepares to hook everything up, he stops and looks at Beth. "I skipped lunch and I'm getting hungry. Can we take a break? I'd like a piece of that KFC."

"No problem. I left the whole afternoon open. I didn't know how long this would take."

Bradley opens the refrigerator and pulls out the tub of chicken. "Take whatever piece you like."

Beth grabs a leg and they toss their selections into the toaster oven on the counter. Bradley grabs the coleslaw and two soft drinks and they pull up stools at the kitchen bar.

"You know you're welcome to visit every Saturday," he whispers with a mischievous smile.

"Brad, don't go there," she insists. The kitchen gets quiet as they wait for the chicken to warm. When the oven dings and Bradley serves up the hot chicken and coleslaw, he breaks the silence.

"You said in class one time you liked KFC. That's why I picked it up today."

Beth waits a minute before she asks, "Have you prayed about our friendship?"

"Actually, I did a few times," Bradley admits.

"So let's consider God has something more in mind than being friends. Let's consider we're together for a special purpose. Perhaps I'll influence your college choice or help you decide which area of science you'll study.

Perhaps I'll start attending your Presbyterian Church and meet Mr. Right." She stops for a minute. "I believe there's something bigger going on we can't see yet."

Bradley bites his chicken and considers her words. "I understand what you're saying." He pauses. "Beth, you know I'm inexperienced with girls, but being with you is . . . Well, . . . it's wonderful. I'm looking forward to that date you promised, but you're right. I should pray about God's purpose."

"Your dad's worried about us. He knows our friendship comes with unique hazards. I don't want to discount our growing companionship, but I'm interested in what God's got planned for us."

"Dad and I prayed about you before we met. Well, we prayed about the new science teacher. I guess I haven't thought about God's intentions."

Beth finishes her meal and looks at Bradley. "We need God to be in control."

"I know you're looking out for me," he says, giving her hand a gentle squeeze. "I appreciate it."

Although enjoying his soft touch, she cuts off the discussion. "Are we ready to get back at it?" she says standing and putting her paper plate in the trash.

Before starting the water, Bradley carefully installs the loops into the channel. He adjusts their placement, turns on the gauge, and releases the water. Watching the speed gauge, he comments, "It's showing less than one mile per hour, but I want a more precise reading."

As he fusses with the voltage, loop installation, and gauge, he says with satisfaction, "I've got one foot per five seconds. That's a more accurate measure."

Looking up, he notices Beth peering into the water channel by the bridge. A small amount of foam is collecting.

"Is there something in the water?" Bradley asks. Beth doesn't reply. She deliberately sticks her finger in the foam and tastes it.

"What are you doing?" Bradley exclaims.

"Stop everything!" Beth shouts, waving her arms. Bradley shuts off the water, the microwave, and the loop voltage. He's never seen Beth so excited.

Beth demands, "I need a microscope, clean glass slides, and an eye-dropper. Do you have those items?"

"Yeah, I think so. Why do you need those? I don't understand what you're . . .?"

Beth interrupts with a serious tone. "I'm transmitting, not discussing."

Bradley gets the message and rustles around the basement looking for his old chemistry set. Opening it, he locates all the requested items. He hurries to the bathroom to clean up the dusty pieces.

"Everything okay?" Conley asks rushing down the stairs.

"Sorry. I'm sorry. I'm getting a little overexcited," Beth counters. She grabs a pill out of her purse and pops it in her mouth.

"Are you all right?" Conley asks as he watches her chew up the pill.

As Bradley comes running back with clean slides, Beth asks Conley, "Can you stay with us for a few minutes?"

"Yes, how can I help?"

Beth motions to Bradley to restart the water. She places a drop of water near the microwave on a slide. She studies it under the microscope and makes notes on a slip of paper. Then she takes a second slide and drops water on it from under the bridge. Following her inspection, and again jotting down notes, she turns to them with an astonished look on her face.

"This exhibit's performing desalination," she exclaims. Conley and Bradley look mystified.

"Bradley, you're cleansing water, purifying, sanitizing, decontaminating," she says, trying to explain the term. There's no response from either Conley or Bradley except looks of confusion.

Beth pauses, calms down, and starts again. "This apparatus has the potential to convert seawater into fresh water. This molecule stretching exhibit, with its microwave heating, nickel plating, and electric gauge, has the capability of making the ocean's water supply available to the world. It could be worth millions of dollars."

Conley quizzes her suspiciously. "How do you know so much about this?"

"The desalination of water is an important technology for climate change resolution. I published a white paper on climate change when I finished my master's degree."

"We've only got a little data. Don't we need more testing?" Bradley questions.

"Oh yes, for sure. I need a more powerful microscope and a comprehensive test plan, but we have a bigger priority." She looks straight at Conley. "We need legal protection. If it turns out I'm right, we must get this invention under patent protection. We can't even exhibit it if it's not fully protected. We need your help."

"I understand, but aren't we jumping the gun?" Conley replies. He's reeling from the twists and turns of Beth's explanation. Her concerned look encourages him to try another approach.

"Look, I have a friend who works for a patent firm in DC. I'll call him on Monday."

"I know I'm out on a limb here, but I'd so appreciate it," Beth sighs. "Bradley and I will put together a test plan next week to verify my suspicions, but we must get legal protection as soon as possible."

Chapter 4

MONDAY MORNING AT 8:30 Conley walks across the hall from his bedroom and reluctantly picks up his office phone. He's wearing his pajamas, but determined to keep a promise.

"Good morning," a polite receptionist says, "You've reached the USP-TO office. The United States Patent and Trademark Office in Germantown, Maryland. How may I help you?"

Conley replies, "I'd like to speak to Mark Conti, please."

"And may I give Mr. Conti your name?"

"Conley Truman. We used to work together in Columbus, Ohio."

"And may I tell Mr. Conti the purpose of your call?"

"Well, yes. I need information about patent applications."

"Have you already submitted an application or are you considering submitting one?"

"I don't know yet," Conley says with a frustrated growl.

"Are you the inventor?"

"No! Can I talk to Mark? Please!" Conley demands.

"I'll put you through," she replies curtly.

After a few digital beeps, Mark picks up the phone. "Hello, Mark Conti."

"Mark, it's Conley Truman."

"Conley, how have you been? I don't think we've talked since Rosemary's funeral. Everything going all right?"

"Mark, it's good to hear your voice. Bradley's working on a science project and his high school teacher thinks he needs a patent. I promised I'd call and put the issue to rest."

"So you don't think his project's novel or unique. Has he invented a new vacuum cleaner?" Mark laughs, and it echoes in the phone.

"He's working on a project called desalination. I know little about it, but his teacher's all excited."

"Is he eighteen yet? If not, I'd suggest you find him a qualified sponsor."

"I guess his sponsor would be his current science teacher. Although the project started under a previous teacher."

"What's the teacher's name?"

"Elizabeth Williams. She's a Stanford graduate."

"Hold on," Mark mutters, "Let me do some digging." Conley can hear the hurried clicks on Mark's keyboard. Within a few minutes he hears Mark's breathing accelerate.

"I thought she sounded familiar," Mark says excitedly. "Her white paper's entitled *Inevitable Climate Change*. My boss required everyone in this office to read it. We specialize in climate change related patents.

"Conley, Williams ranked number two at Cornell and number one at Stanford. One commentator wrote, and I'm reading from his comments on her paper, 'If she ever grows up she'll be the leading innovator and trouble-shooter for US climate issues'. Are you sure we're talking about the same person?"

After a short pause and many more keyboard strokes, Mark says, "I'm looking at a year book picture of her at Stanford. She's about five foot six, short dark hair, unusually fair complexion, and looks about fourteen years old."

"That's her. She's Bradley's teacher."

"Then let's get down to business. I won't doubt her judgment. I'm telling you that desalination know-how is in demand everywhere. Because of climate change ocean forecasts, there's a world-wide race for desalination dominance. Controlling desalination's very political and somewhat danger-ous. There are global factions that'll do anything to gain the technology.

"After the South African incident last December, security's been stepped up at our offices. Our new receptionist's both an FBI profiler and markswom-an. It's her job to ensure the safety of our patent documentation. Our depart-ment even installed a customized bank vault to protect our patent filings."

Conley responds, "I know nothing about South Africa, but okay, I'm a believer. What should be our next steps?"

For the next thirty minutes, Mark provides Conley with a step-by-step process to complete the patent filing. The first step is getting signed non-disclosures. The next step, is a detailed description of the invention. Once Mark's office has checked to confirm that no one else is using the same process, they'll prepare a patent application. Mark stresses the need for secrecy until a provisional patent's issued. Conley takes two pages of notes as he listens.

"Is there anyone besides you, Bradley, and Williams who understands what Bradley's invention does?" Mark asks.

"I suppose his previous science teacher, Mr. Galloway," Conley answers.

"I know you can prepare a tight non-disclosure," Mark urges. "Make this one a doozy. Mr. Galloway, you three, and anyone else who knows what Bradley's doing should sign it as soon as possible. Set up a documentation file and keep it in a safe place. For your patent application, I'll consider Bradly the sole inventor, Williams the sponsor, and you the legal advisor. Anything else?"

"Mark, thanks so much. I never realized this was such a big deal."

"Welcome to my world," Mark chuckles. "It's been great talking with you."

Conley's already thinking ahead regarding the patent steps. He hangs up the phone and checks the time. He dresses for the day and drives off in his F-150.

"Good morning," Conley says to the high school principal's admin. "Any chance I can visit with Mr. Maxwell for a few minutes?"

She cracks open Jordan Maxwell's office door and mentions Conley's name. Soon Principle Maxwell comes out with his hand extended.

"Conley, this is a pleasure. What brings you to our high school? I hope there's no problem with Bradley." Mr. Maxwell motions for Conley to follow him into his private office.

"Jordan, I've an unusual request. I want to know if I can invite Ms. Williams to move into my house for a couple of weeks."

"Well, that is unusual. Doesn't she have her own place? Did she ask to move in with you?"

"I haven't discussed it with her. I wanted to be sure I'm not violating any school policies before I extend the invitation."

"Do you mind if I ask why?"

"Not at all. I'm working on a legal patent that includes a lot of scientific terminology and processes. I'm aware of Ms. Williams' education and thought she could be an invaluable resource. This assignment is under a tight time crunch, so it just seems more practical if she stays at my home. It would make everything easier for both of us. What do you think?"

"There's no specific policy against what you're proposing. But you can't pressure her, and you must be careful of her involvement with Bradley. Am I right?"

"Absolutely, I couldn't agree more," Conley nods seriously.

"It's important her assistance doesn't interfere with her teaching. You're not planning to pay her, are you? I mean make her an employee or a private contractor. That would violate policies regarding teachers."

"No. I don't plan to give her a formal wage. I thought a reasonable gift might be appropriate."

Jordan stands. "You're a lawyer. You understand these things perfectly. If she agrees, have Ms. Williams contact me and we should be fine." He pauses.

"Conley, a few rumors are floating around about Brad and Ms. Williams. I don't believe any of it, but if she moves into your home, her relationship with Bradley may cause . . . Well, . . . more unpleasant talk. She needs to understand the risk."

"Yes, you're right. I couldn't agree more," Conley states for a second time. "Thank you for your time. I appreciate your understanding."

Leaving Maxwell's office, Conley checks with the admin. "What time is lunch? I'd like to see my son, but I don't want to interfere with his classes."

"Lunch is in thirty minutes. Can I tell him you'll meet him here?"

"I think he's in science class now. That's where I'll meet him, if that's okay. I know how to find it."

The admin prepares a note for Ms. Williams and sends it with a student. Conley goes to the school library and reviews the notes from his call with Mark.

Just before the lunch bell, someone knocks on the science room door. A student hands Ms. Williams the note.

"Class, there'll be a pop quiz on this chapter tomorrow. Mr. Truman, please wait after class. Your dad will meet you here." The bell rings, and as the classroom empties, Conley waits at the door.

Stepping in the room, Ms. Williams notices his serious demeanor. "You've been on the phone with USPTO."

"What's a USPTO?" Bradley questions.

"The United States Patent and Trademark Office," Conley responds.

Ms. Williams closes the science room door and the three find seats. Conley pulls out his notes and says, "We're about to board a fast-moving train. It means a lot of work in a very short time. Are we sure this is what we want to do?"

Bradley nods "yes" with a thumbs up motion. Ms. Williams looks at Conley's somber posture. She's appeals to him as she asks, "Conley, can we count on your full support? Bradley and I cannot do this alone."

"If I'm honest, I'd prefer to avoid this kind of intense schedule, but I'll do it for my son. I'm in 110 percent."

"Thank you," she smiles. "I'll admit I'm uneasy about the schedule, but this challenge makes me wonder if it's God's reason I'm here. Let's go for the patent."

Conley nods. After carefully summarizing his patent notes, he sticks them in his shirt pocket.

"From now on, the desalination project must remain secret. We cannot reveal it to anyone until we have an approved provisional patent application. Each of us must sign a non-disclosure agreement, an NDA, regarding secrecy. The NDA also requires the signer to relinquish any un-authorized rights to any part of the invention. This NDA applies to any person who understands its purpose. I'm thinking of Mr. Galloway, but is there anyone else?"

"What about the guys who built the metal frame?" Bradley asks.

"Do they know about the science?" Conley replies.

"Well, I went over the water flow in great detail. I wanted them to get it right. They know nothing about desalination."

"To be safe, we should ask each one of them to sign. While you're there, try to acquire their CAD drawings or whatever written information they have."

"What about the kid who painted the apparatus?" Bradley questions.

"Again, what did you tell him?"

"I told him the water starts here, travels through there, and ends up in there." Bradley's arms are motioning as he pretends to be in front of the frame.

"No. I don't think we need his signature. Anyway, his signature's meaningless if he's a minor."

Conley pulls the notes out of his pocket.

"This week," he says, "we need to summarize the project from its first thought to its present development. Whatever documentation can be created, no matter how high-level, we must ship overnight to Mark at the USPTO with a copy mailed to our home address.

"The following week, the summary must dig deeper with questions like: When did this project begin? Where did this idea come from? Why did we pursue it? What were the results? All that information must accurately describe the water stretching exhibit. Then we will provide that summary again to the USPTO and ourselves."

"Water stretching?" Bradley states. "This is no longer about water stretching."

"Details, details, details," Conley responds. "This project has changed, but the original scope's been integral to our current discovery. To make this invention patentable, the USPTO team needs to understand every step.

"We'll keep following our original summary and break this exhibit down until we've given my contact, Mark Conti, and his team every detail."

For a moment nobody speaks. Bradley and Beth's brains are churning through everything Conley's identified as urgent tasks.

"Dad, you said we had to sign away our rights to the project. I don't understand."

"Everyone who signs the NDA will give up all rights. The patent, however, will state that you're the inventor, Beth's the sponsor, and I'm the legal representative."

Turning to Beth, Conley speaks in a serious tone. "Beth, because of the time restraint, would you consider moving into our home for a couple of weeks?"

"I can see why that makes sense. We've got a lot of research to coordinate and complete. I'll discuss moving with Principle Maxwell."

"Please don't be angry, but I've already discussed it with him. I didn't even want to ask you, if it violated school policy. Even so, you'll still need to finalize with him." Conley pauses for a moment. "If you move in, you need to consider there may be some reaction from students and teachers. Mr. Maxwell believes your friendship with Bradley is under a microscope."

"If we can do this in two weeks," she replies, "I'll be back in my apartment before any new gossip can get started."

The bell rings, signaling lunch time is over. The three stand. They look at each other with nothing more to say.

Before walking out the door, Conley suggests, "Let's have dinner tonight. We can talk this through again. We'll have something simple. Is

5:30 good?" Beth shakes her head with agreement as each treks off to their afternoon schedules.

That evening, dinner conversation's a free-for-all. Maybe it's because they unanimously agreed on BLTs that they can't agree on anything else. All three argue for different priorities to complete the patent prep. Conley's concerned about signing the NDAs, Beth's eager to test for desalination substantiation, and Bradley wants help to find his notes. After forty-five minutes of friendly arguing, they're frustrated and exhausted.

At last, Beth breaks the impasse. "It's obvious our immediate interests do not coincide. We must focus on the legal process at this stage, not the scientific one. We'll never get the science finished without settling the legal. Here's what I propose: first, I move into your home; second, Conley helps us complete as many NDA signatures as we can; and third, Bradley and I work together on finding and summarizing the data we have. Can we do that this week and target a mailing on Saturday?"

"Yes," Conley sighs with relief.

"I like the first step the best," Bradley teases, "but tonight's discussion shows this is a five-star disaster."

"I agree," Conley laughs. "A five-star disaster calls for specific actions." Conley looks at Beth with a mysterious grin. "You in?"

"Is this an inside joke?" she queries. "I don't know what to say."

Bradley grins. "Specific actions start with prayer . . . and end with ice cream sundaes."

"Oh, I'm in!" she beams.

They bow their heads and each one prays for the success of their patent preparation steps and for each other. When finished, they motor to their favorite Jim Dandy restaurant.

All week, the three work at their agreed upon priorities. Beth moves a few essentials into the guest room, and Bradley puts up a cot in a storage room off the basement. Conley prepares a tight non-disclosure agreement and opens a large safe deposit box at their bank. Bradley sorts through dozens of papers finding odds and ends of his research notes for Beth to summarize.

During the week, Beth and Bradley visit Mr. Galloway. He willingly signs an NDA and wishes Bradley a favorable patent outcome. He also provides several personal notes on Bradley's initial exhibit discussions. Before they leave, he informs Beth of the school's science data storage system, which may have more useful information.

On Friday night, Beth puts the finishing touches on the first USPTO summary and prepares the mailing. Since the summary's missing detailed descriptions, she encloses photos of the current exhibit setup. She also provides copies of the signed NDAs.

Saturday morning, everyone sleeps late. At breakfast no one looks ready for the day. "What are your plans today after mailing our summary?" Bradley asks Beth over coffee.

"I'm going home and pick up more clothes and personal items," Beth responds.

"I'm taking these NDAs to IronWorks. Depending on their response, I might be quite a while. We can catch up later this afternoon."

Arriving at IronWorks, Inc. fabricating shop, the owner greets Bradley in the front office.

"Tell me, did we 'wow' the senior class with our welding?" he shouts. The owner, Fred Collsman, graduated from Springfield High about five years ago. Bradley remembers him as a loud mouth motor head, but talented with a welding torch. After graduation, Fred joined his dad at Iron-Works, Inc. It was a well-known place for custom metal structures. When Fred's dad died from an unexpected heart attack, he took over the business.

"Not yet, the school's art and science exhibit is not till the twenty fourth," Bradley replies. He opens a manila folder and pulls out a half dozen NDAs. "I'm asking you and your employees for special cooperation on the exhibit framework you built for me."

The owner calls to the three employees who worked on Bradley's metal frame and then picks up the NDAs.

"What are these?" Fred says mockingly, waving Bradley's agreements in the air.

"I'm applying for a patent on the molecular process that's exhibited on the frame you built. The patent office requires formal secrecy, so my dad has prepared these agreements."

He passes an NDA to each employee. "Since I'm providing free advertising for IronWorks at my exhibits, I'd like you to sign these agreements. Please give me any diagrams or drawings you created."

"How much did you pay us for that work?" Fred asks.

"I think the bill came to just under four thousand dollars," Bradley remembers.

"Sounds right. That was a discount considering your lawyer dad helped us get the business into a C-Corp. But that price didn't include signing any papers or giving you our drawings. That'll be extra, and I mean extra for all of us." Fred laughs and his employees grin.

One employee asks, "What am I signing, anyway?"

Bradley keeps his voice firm and professional. "By signing the document, you agree to keep any knowledge of my molecular science process a secret. You also give up the right to receive any benefit from its design or use should it become marketable. Except for admitting you designed and built the frame, you're agreeing not to say anything about its purpose."

Another employee speaks up, "So what happens if you sell your process on our frame for a million dollars?"

"You'll have no right to any of that money," Bradley responds.

"So what are the chances you'll make a million dollars?" Fred grins.

"Until I get a patent, I've no idea. I don't even know if I'm eligible for a patent. Right now I'm just trying to keep the idea a secret."

Fred winks at his employees. "I think cash today is better than a lottery ticket for tomorrow. I'm thinking Bradley should pay one thousand dollars in cash to each one who signs. How does that sound boys?" Two of the employees grab a pen and sign before Bradley can respond.

Fred sweetens the deal. "If everyone signs, I'll throw in our diagrams for nothing."

For the first time, Bradley's voice sounds distressed. "Okay," he sighs. "If I can get everything I need today, all the signed NDAs and the CAD diagrams, then it's a deal." Everyone nods their heads and Bradley turns to leave.

"Where are you going?" Fred asks.

"To the bank. I don't have that much cash on me. I'll be back in about thirty minutes. Make sure you read the agreements."

Bradley goes to the bank, withdraws cash from his savings, and pays Fred and the IronWorks' employees. Back home by 3:30 p.m., he tosses the signed paperwork on his father's desk knowing he's broken a promise.

Before she died, Rosemary stressed Bradley should save money for a car. She said Dad will cover college tuition, but he must save and buy his own vehicle. For the last five years he's put money away: skipped dating, skipped ball games, skipped movies, and even avoided hanging out with the boys. Every cent he's earned went towards his car fund. Now his invention has depleted more than half his savings. He's exhausted from the week and discouraged from tapping into his savings. Making his way to his lower level bedroom, he lies down to sleep away his misery.

Chapter 5

BRADLEY OPENS HIS EYES and sees Beth sitting cross-legged on the floor. When she realizes he's awake, she pulls herself up on her knees, leans forward, and lays her head on his chest. She strokes his blond curls with her fingers.

"This week's been hard," she whispers. "This project has overextended us. I know I've been hard to get along with: crabby, impatient, and exhausted. When I walked into my apartment, my first thought was, 'I wish I'd never gotten involved'.

"Then I realized how much I care for you. I remembered we prayed together for God's purposes in our lives. When I thought about how hard you've worked to get your exhibit completed and the money you've spent, I felt so selfish. I laid down and cried myself to sleep.

"Brad, I don't know what God wants, but whatever it is, I believe he'll give us the strength to finish it. If we can survive next week and complete the patent details, maybe we can catch a break. But let me say, living here has been . . . delightful.

"I wish we could sit together in a secluded place. A place where we could gaze at the stars and city lights. Your arm around me and my head on your shoulder, we don't even need to talk. Just you and I together, away from all this craziness."

Bradley listens, but doesn't know how to respond. Surprised and pleased by her show of emotion, he relishes her tenderness, the smell of her hair, and her hand in his curls. He wants to say 'I love you', but knows it's not the time. He must not weaken his resolve to protect her reputation. Beth slowly slips out of the room; he says nothing. His melancholy about his savings has evaporated. He'd gladly spend twice as much for another intimate moment with her.

On the twenty fourth, the Springfield High gym looks like a circus at 4:00 p.m. Students and teachers are busy setting up for the Annual Art and Science Fair.

Bradley backs the F-150 up to the outside gym doors. The pieces of his exhibit framework are in the six-foot pickup bed. After unloading, he pushes the frame segments over to the science side of the gym. Not sure where to position his framework, he asks for help.

"Coach Jefferies, do you know where the seniors are setting up their science exhibits?" Bradley asks the basketball coach.

"I hate what this does to the gym floor every year," the Coach whines. "It's already full of deep scratches and the public hasn't even arrived. Sorry, Brad, the senior area is over there." He points to a section where Bradley recognizes Linda and Terry setting up their exhibits.

Bradley pushes his framework alongside Terry.

"Your exhibit's in the wrong place. Shouldn't you be in the art section?" Terry says with sarcasm.

Linda comes to his rescue. "Don't pay any attention to Terry. Are you going to show the water stretching you talked about with me?"

Bradley doesn't respond. His mind flies into overdrive as he tries to remember his conversation with Linda. *And what about Karen?* he thinks. She worked on his computer program. Bradley panics, wondering if he needs two more signed NDAs.

While pinning the IronWorks' frame together, his eyes are searching everywhere for Beth. Seeing her come in from the parking lot, he calms down and completes the frame assembly. About the time he's finished, Beth comes to check on him.

"Are you doing all right?" she asks.

"It occurs to me that Linda and Karen helped create my exhibit. Should I have them sign NDAs? I hope I didn't miss a step," Bradley says apologetically.

"Take a breath," she laughs. "After tonight's exhibit everyone will know what you're doing. Will Linda and Karen know more than others?"

"Well, Linda won't, but Karen designed the program I'm displaying on the monitor."

"I'll talk with Karen. You need to concentrate on getting ready. This exhibition starts soon."

By 6:30 p.m. the gym's packed with adults, parents and friends, who plan to review the artwork and science exhibits of the high school kids. Conley arrives and spends a little time wandering around before joining Bradley.

"Conley," a familiar voice calls, "I thought you'd be here."

"Mr. Maxwell," Conley replies. "Are you enjoying the work of your students?"

"Well, I'm always impressed with our young talent. Tell me, is Ms. Williams still staying at your home?"

"No. She moved back to her apartment last weekend. Two weeks was enough time to complete my patent project. Thanks again for your cooperation."

"So far tonight I've talked with five different parents about Ms. Williams. Each one says that in the last two weeks, their kids have come home excited about science. Ms. Williams seems to have a new energy. She's taken off like a rocket while staying with you. I hope that's a good thing."

"Well, it sounds good. Do you agree?"

"To be honest, I'm worried she spends too much time with two bachelors, if you get my meaning."

"When Ms. Williams completed her master's degree at Stanford," Conley says in a matter of fact reply, "she wrote a white paper on climate change. While assisting me in understanding scientific patents, we learned her paper's required reading for certain US government employees. I think it gave her a real shot in the arm."

"I'm so glad you said that," Jordan replies. "Just between us, at my last school assignment a student and teacher ran off together mid-year. The parents, the press, the paperwork . . ." Mr. Maxwell sighs and shakes his head. "It was terrible. I wouldn't want that situation again."

"You've nothing to worry about," Conley comforts him. "She's back in her own apartment busy with her many students."

"Dad," Bradley calls. "Come here. The water's up to twelve centimeters. I reached my goal tonight."

Conley and Jordan hasten over to his exhibit. A stream of water is flowing up over the bridge, and the metric scale's pushed to the maximum.

"I hope Brad's going to enter the state competition," Mr. Maxwell says. "This looks like a great exhibit. It's always a feather in my cap when one of our students does well at state."

At 9:00 the gym lights blink three times, a signal to conclude. As parents and friends exit the gym, Beth alerts the senior science exhibitors they may dismantle their projects the next morning. Conley and Bradley drain the channel and unplug the electronics. As the gym clears, the janitorial staff starts to clean up.

An hour later, an older European man enters the gym. He's escorted by the janitor who leads him straight to Bradley's exhibit. The European meticulously studies the framework, takes several pictures, and hands the janitor a padded envelope. As he exits the building, the janitor notices a silver hearing aid on his right ear.

The next Saturday, Beth and Bradley prepare to measure the desalination quality. Beth borrows an LED binocular compound microscope from the local community college. Bradley picks up a ten-pound bag of seawater salt from a nearby pet store. Their challenge is to compare water droplets from various locations in the channel. They hope to find evidence that saline parts per million decrease from the channel inlet to the bridge.

"Where's your dad," Beth asks after they carry all the equipment to the basement. "I notice the LTD's missing."

"He'll be back soon. He's running a few errands. Not sure where he went," Bradley replies as he leans against the basement wall.

"Beth, about that day." Bradley points to the storage room. The place where he slept when Beth stayed with them. "I'm sorry I didn't respond to you."

Beth sets the microscope on a shelf and looks him in the eyes. He takes her hand.

He speaks slowly, "What I wanted to say, I didn't say. But it was a very special moment for me." Just then they hear Conley enter the kitchen door. They at once drop their hands, but that short touch says everything they wish to put into words.

Conley comes down the stairs with three coffees. "What can I do?" he says.

"We need someone to record our readings," Beth suggests. "I brought a yellow pad. If you don't mind taking notes, that would be very helpful."

"No problem, let's go through it once for my unscientific mind," Conley laughs.

"So Dad, this test is our control exercise. We're using Springfield tap water, and our intention is to stretch water over the bridge. We'll be taking water samples from various channel points and asking you to record them."

"Got it," Conley answers, finding the only folding chair in the basement.

Bradley starts the water and Beth puts water samples on slides for the microscope. She reads the results out loud. "Conley, the saline parts per million, what I mean by ppms, of the tap water is 515. After it flows over the nickel catalyst, it's about the same, 487 ppms. At the water next to the bridge, it's 453 ppms. The same at the collection tank, 453 ppms. Let's do it again."

For the next hour, they perform the test on the tap water over and over. Each time Conley records the results, and each time the results are slightly different. After ten identical tests, Beth advises Bradley to shut off the water.

"I'm looking at our ppm numbers," she says reviewing Conley's notes. "The tap water range measures from 492 to 519 ppms, the nickel catalyst ranges from 479 to 495 ppms, the water at the bridge measures between 447 and 459 ppms, and the collection tank pipe's 445 to 458 ppms. Next we'll test saltwater."

She puts down Conley's notes and says, "Anybody hungry?"

Bradley and Conley look surprised.

"Well, it's 11:50," Beth responds. "We need to top off the tank with water and add the saltwater pellets. It'll take at least an hour for the pellets to saturate the water tank."

"If you two want to run for a bite to eat, go ahead," Conley states. "I've got to check on several matters in my office. I'll meet you back here in about ninety minutes."

"Come on, Brad, we'll take my car," Beth says as she grabs her coat.

Walking out the front door, Bradley asks, "Where can we eat without being seen together?"

"Don't worry," Beth replies, stepping into her Chevy Geo. "We'll eat in the car after ordering at a fast food. I know just the place."

Beth maneuvers the winter roads to a strip of commercial food restaurants near the interstate. "Have you ever eaten at Smalley's Tavern?"

"Never heard of the place," Bradley responds.

"Then I'll order for both of us and you'll pay." Beth laughs and pulls into Smalley's drive through line.

At the order window, she requests two turkey and roast beef double-decker club sandwiches with all the trimmings. She also orders two hot grande mocha lattes. After Bradley pays, she hands him their meals so she can drive into a nearby abandoned department store parking lot.

"We'll have privacy here," she says. Parking the car, she pushes the driver's seat all the way back.

At first they're too busy eating to talk, but Beth's waiting to confess.

"Your dad told me how much the IronWorks frame cost you. I feel like it's my fault."

"Why?"

"Well, that day when I poked fun at your original contraption, I asked you to make something more presentable. I never meant for you to spend four thousand dollars. Then the NDAs cost another four thousand." Beth takes another sip of coffee, but closely watches Bradley's expression.

"Before my mom died, I made a bunch of promises to her. One promise was to save money for a car so Dad wouldn't have to support both college tuition and my transportation. I stayed away from using any of my income for dating, school functions, or anything for myself. When I went to IronWorks, I broke that promise."

"I'm so sorry. Why did you do it?"

"Because of another promise I'd made to Mom. I promised when I met the girl of my dreams I would do everything in my power to honor her." Bradley's head drops. "I'm sorry, you didn't need to hear all that."

"There's nothing to be sorry about. You're an honest guy and I love that about you," she replies stroking his hair.

"The day you laid your head on my chest," he says, "I felt depressed because I broke my savings promise. Your tenderness, however, reminded me I hadn't broken all my promises."

For the next few moments, they sit in silence. Each one caught up in their own thoughts.

"I wish my dad made such a promise to my mother. Perhaps they would still be together," Beth murmurs. "I think we should head back." Bradley nods and in a few minutes they're traveling back for the afternoon testing.

As Beth drives, Bradley asks, "What's the pill you took last weekend? Dad said you grabbed a pill from your purse."

"So, are you two checking up on me?" she laughs. "It was just a gummy for high acid. When I get emotional, my stomach gives me heartburn. I don't need it for you though."

Conley's in the basement when they arrive. He's brought his work laptop and is busy entering data from the morning results on a spreadsheet.

"Hey," he says, "I need just a couple more minutes."

"No problem," Beth replies. "Let's test the saline concentration."

Bradley starts the water for a few seconds and Beth puts a drop on a microscope slide.

"It looks like the ppms are about thirty thousand. That's not quite seawater level, but close. The ocean averages thirty-five thousand ppms, but this water is perfect to test."

For the next two hours, the three collect desalination data. The results are very positive, but after test number four, they take a short break. Bradley gets a bucket from the garage and places it under the channel to catch the overflowing foam. Once six more tests of the home-made seawater are finished, Conley arranges the results into a proper format.

He reads from his spreadsheet. "Saltwater range measures from 29,985 to 31,087 ppms, the nickel catalyst ranges from 29,857 to 31,064 ppms, beside the bridge measures 786 to 852 ppms, and the collection tank pipe is 785 to 852."

Conley puts his laptop down and extends his arms for a celebrative group hug.

"We're on the right track," Beth exclaims. "This project's a winner. Something about those twelve volt loops causes the saline partials to drop."

Bradley's thrilled. "We'll knock the socks off the state judges with a patent approval."

Beth nods her head. "On Monday we can send our desalination results to Mr. Conti. This is so exciting."

Picking up her coat and notepad, she heads towards the door. "Well, I should go home, take a nap, and write up another test plan." She turns to Bradley. "Thanks for your openness today. It meant a lot to me. See you Monday."

Chapter 6

THE EUROPEAN CLIMATE CHANGE meeting is held the second Sunday in March. At 9:00 a.m. President Diego Gonzalez steps to the EU headquarters conference room podium. Before speaking, he taps the microphone. Upon hearing the sound system answer back, he speaks in English with a strong Spanish accent.

"Welcome ladies and gentlemen of the European Union. I'm Diego Gonzalez and I serve as president of the Eurogroup. It's a pleasure to have all twenty-seven of our EU countries represented in this conference hall. If you've counted the chairs, you know we planned for about seventy-five attendees. We invited each EU country to send two qualified representatives and sent courtesy invitations to our NATO friends and non-EU European neighbors.

"Today's meeting is the preparation for the Eastern Hemisphere Climate Change Symposium the end of May. The EHCCS will be a panel discussion format to discuss climate change contingencies for all Eastern Hemisphere countries. This meeting's emphasis, however, is to consider what's best for Europe and the European Union in particular." The attendees clap with excitement.

"Today's agenda is straightforward. After a few brief guest introductions, we'll break into small groups of eight to twelve and discuss four climate change predictions. After discussion, each group will recommend specific actions regarding each prediction. Once we've completed the group sessions, we'll take a lunch break before summarizing each recommendation. Before leaving today, we'll vote to determine the top proposals and their related activities. Just so no one misunderstands, only EU members are eligible to vote."

Motioning to someone on the front row, President Gonzalez says, "Now let me introduce the leader of Eau Suprême."

An attractive Italian woman in her late forties walks to the platform. As she lifts her head to address the audience, several individuals clap.

"Thank you. As many of you know, I'm Alessia Amato. It's my privilege to lead Eau Suprême."

Ms. Amato adjusts the microphone to accommodate her almost six-foot frame. "Eau Suprême's primarily organized to protect Europe's fresh water resources and related businesses. The EU created this organization because projected climate changes will impact fresh water supplies.

"Since its creation, Eau Suprême has expanded to incorporate growing political concerns considering both foreign immigration and military aggression. The EU recognizes that our rich supply of fresh water could become a target for populations lacking a sufficient supply.

"I'll agree with many who claim hostile conduct will never transpire. But until proven otherwise, it's my job to plan contingencies for every misfortune possibly affecting Europe's fresh water resources. Thank you."

As Ms. Amato steps down, the president says, "I've one more person to introduce before we break into groups. Dr. Edwin Martin has flown from Australia to give a brief report on a promising research project."

A large, casually dressed man sitting beside Ms. Amato stands and hastens to the platform. He turns towards the audience with a large grin.

"G'day," he says laughing. "Ow ya doin' mate." Then, with no further trace of an Australian accent, he speaks in the King's English with a strong hint of Scottish Highlands. "I didn't want anyone to think I wasn't in Australia," he remarks with a smile, "but my home is Edinburgh."

Then, with a serious tone he declares, "The greatest criminal on planet earth is carbon dioxide. In our industrial world, humanity is guilty of breeding CO_2 in every developed country. The only natural consumer of CO_2 is the plant kingdom through photosynthesis.

"Right now at the Australian National University, a dozen world renowned chemists are working towards developing artificial photosynthesis. We've made several breakthroughs, but our target is still around the corner. While you're busy protecting our great European resources, the Australian team hopes to bring CO_2 under control with new photosynthesis technology. I'll be here all afternoon if you have questions. Thank you."

President Gonzalez comes back to the podium. Before speaking, he passes questionnaires to several staff members and gives instructions for the group sessions.

"We are asking for your consensus on four areas of climate change. These areas should be no surprise as I read you the list:

1. Extreme changes in heat and cold.

2. Severe and damaging weather.

3. Loss of coastal shorelines and cities.

4. Insufficient fresh water.

"These four areas are listed at the top of your questionnaire in no specific order of priority. Each group may choose its own priority to start discussions, but try to complete all four before our break at noon."

Motioning for the staff to distribute the questionnaires he continues, "As you form your groups, it's important that discussion includes a cross section of our diverse country representatives. I suggest that each group comprises delegates from at least four different countries. Please choose someone to record your discussion highlights as well as a spokesperson. Besides collecting your written notes, we'll ask each spokesperson to express your top recommendation for Europe.

"Thank you. You may form your groups now. You're welcome to rearrange the chairs as required. Remember groups can be as small as eight, but we suggest no larger than twelve."

For the next few minutes there's a lot of chatter and thumping of padded chairs, but soon everyone settles down.

Vice President Kennedy Vincent Kleaver, one of the two US attendees, leaves the room and strolls over to Ms. Amato's office. Knocking on her glass entrance door, she motions him to enter.

As she greets him with a hug, he says, "Alessia, I'm very pleased to represent the US in this work session. I hope we get several valuable action items by the end of the day."

"Mr. Vice President, I've known you too long. You didn't come to my office for pleasantries," she responds, "and I'm all out of Belgium chocolates. What's really on your mind?"

"Do you remember a Swiss German bottling company called Wonderful Water Intake or something like that?"

"The name is properly translated to English as Wonderful Fluid Intake. Their Swiss German name was Wunderbare Flüssigkeitszufuhr. They went out of business several years ago. Why the interest?"

"If I remember correctly, their CEO was a politician wannabe. The real force behind the company was the CFO, a man named Finn Schweitzer. What ever happened to him?"

"You wouldn't ask me that question, if you don't already know the answer," she laughs, "but Finn's the Swiss Minister of Environment, Transport, Energy, and Communications. Actually, he's assigned to me as part of Eau Suprême. Come on, Ken, I'll bet you know all this."

"We're suspicious Finn had a hand in the desalination plant disaster in Nhlabane, South Africa. We've no solid proof, but a ship guard in South Africa identified one of his old comrades, a Karl Muller, as leaving a cargo ship in the Port of Richards Bay. The ship guard spotted Karl the same night as the plant violence, and he never saw Karl return to the ship. In addition, a flight attendant at the King Shaka International Airport identified Karl as a passenger on an early morning flight to Zurich the next day."

Alessia replies, "I'm sure Finn has moved away from the mischief of his water bottling days. That bunch of criminals who worked for him almost ruined his career. A few went to prison. I think you're on the wrong track. Maybe this Karl person is working on his own. Now, tell me about your family."

As Kleaver shares the joys of his growing number of grandchildren, the conversation drifts back to their Peace Corps days in middle Africa. For two years, they worked with international teams. The members became close and remain lifelong friends.

"Have you seen or heard from Frank?" she asks.

"No, Frank's not happy with me. Perhaps someday I can make up for all the trouble I caused him and his family."

Alessia's admin sticks her head in the door. Speaking in broken English she announces, "The group discussion's finished. They are breaking for lunch."

"Thank you, Lina. You remember Vice President Kleaver, right?"

Lina nods and holds the office door open. As they walk towards the EU cafeteria, the VP gives a discreet wink to Lina.

The climate change meeting resumes after lunch.

"All the notes from this morning's discussions have been collected. Our staff is preparing the information for voting later this afternoon. In the meantime, we want to hear from our group spokespersons.

"Let's start with proposals on extreme changes in heat and cold," Gonzalez states. "How many groups chose this topic for their top recommendation?"

Out of eight groups, only one person raises a hand. The spokesperson stands and speaks extemporaneously. "Our primary concern is the high heat. Most northern Europeans know how to handle bitter cold, but in many countries it's not common to have air conditioning in homes or public buildings. Our recommendation is that each country select suitable public buildings, particularly schools and theaters, for air conditioning installations."

"Thank you. The next one we'll consider is severe and damaging weather. Who's chosen this subject as their top priority?" Two hands raise and President Gonzalez asks them to speak one after the other.

After giving their reports, Gonzalez continues, "What about loss of coastal shorelines and cities?" No one raises a hand. The president's surprised and asks again. "Doesn't any group consider the loss of our coastlines as a top concern?" No one responds.

"That leaves us to discuss insufficient fresh water supplies. Have the five remaining groups prepared recommendations for that topic?" Five hands raise.

The President acknowledges a woman in the back. She stands and voices her group's recommendation. "President Gonzalez, our group concern is with the potential lack of water for Europeans. We realize many of our countries have unlimited water resources. However, once we supply our arid neighbors, we may find ourselves without water. We recommend that each country place an applicable limit on water exportation. Our intention is keeping enough fresh water within our own borders."

Immediately another group leader stands and announces, "With all due respect to our friend's recommendation, our thinking's quite contrary. We think exportation of water's a strong business for Europe. Our recommendation is to increase bottled water distribution, especially to arid regions, primarily for profit. Let us drink our wine and get a return on our water resources." The hall fills with laughter.

"Thank you for those interesting and opposite points of view," says Gonzalez. "Who's next?"

As President Gonzalez conducts the afternoon session, another less formal meeting's taking place in an abandoned property near the Bern Airport. The old hotel is for sale. The accommodations appear functional, but there's no hotel staff. Finn arranged the secret meeting place through Stefano and one of the Solé water company branch offices.

A little over a dozen men, loyal to Finn Schweitzer, meet in the dusty dining room.

"Friends, it's so good to see you again," Finn begins. "I'm happy to say the success of our operation continues. I thank you all for that. We hope to turn the bottled water industry to our way of thinking by the end of the year. Let's stop this ridiculous and unnecessary conversion of seawater and make a profit on the water resources God has given our countries.

"Regarding Nhlabane, no one's figured out we sabotaged the diesel engines. In fact, the deaths are still unsolved. We can give credit to Karl for that. In California, our colleague, Hans, has infiltrated a target for over five months. Our plan is closing that plant where a revolutionary steam process is being tested. We've successfully gotten our Greek friend, Alec, hired at a plant in Ras Al Khair. We've also installed colleagues at several other locations where their skills allow them strategic positions.

"I realize it's risky meeting this way, but I want to make you aware of how we're stepping up our plan. The EHCCS in May will be our first event to hit global news. We intend to put a scare in the hearts of desalination promoters. If we're able to focus the damage, the reports will blame the event on Alessia. As you all know, I want her job. Once in her position, I can freely endorse and support our bottled water companies. Are there any questions?"

A man at a nearby table speaks, "Finn, can we be sure Alessia's not aware of our actions?"

"If she is, what can she do? Now, over 65 percent of bottled water companies have executives loyal to our cause. The water industry'll stop her unless she's craftier than I think. We need her out of the way," Finn says angrily. He pauses and smiles. "I don't think she knows anything. I've a loyal supporter working in her office." The whole room chuckles at the irony.

An older German man stands to speak. Out of respect he removes his cap revealing two noticeable features: a silver hearing aid in one ear and a small patch of black hair on his otherwise gray scalp. "You've assigned me to examine new technology in the states. My partner and I have stumbled across a California company which resells brine. Frankly, I can't imagine such a thing. May I suggest we give it a closer inspection?"

"I agree," Finn nods. "I can't see anyone making a living selling brine, but we need to crush the idea before it takes hold. Fred, we'll talk in a few minutes, but I want you to continue your good work. I'll send someone else to infiltrate the brine company. Anything else?"

No one responds.

"Okay. Let's dig into these Swiss treats and German beer. Beer made with fresh European water. I'm glad you risked coming. It's been a long time since Wunderbare Flüssigkeitszufuhr. We'll talk more about our plans as the evening progresses."

The vice president's plane sits on the BRU tarmac outside of Brussels. He's reviewing the conference proposals, including the final votes on each topic. He's thinking the trip wasn't a total waste, but he's sure he could have guessed these results without a trip across the pond. One final phone call to make before flight. He uses his middle name for security.

"Hello, Vince here," he says.

"This is Lina," Alessia's admin responds.

"Are you holding up all right? Do you need anything?"

"I'm fine, really. Do you get my emails?"

"Yes. The secure network's working very well. Are you still hearing from our German friend?"

"It was once a week, but now less often. He asks the same questions about Alessia's itinerary and priorities."

"Anything new at all?"

"He recently asked about Hans Guttmann, but his name's not crossed my desk."

"I'll check him out. As you know, we won't be able to talk again until May. Email me if there's any trouble for you or Alessia."

"Thank you."

The phone line drops and Vice President Kleaver connects his seat belt and sits up in his chair. He's sensing something's not right. He's never trusted Finn Schweitzer since Finn wiggled his way out of a twenty-five-year sentence almost as many years ago. The courts found Finn and his gang of employees guilty of several crimes, but now he's a Swiss minister. Kleaver wonders how that happened, but when he gets back to DC, he'll track down this Hans Guttmann.

Chapter 7

KYLE KRANZTON PULLS HIMSELF out of bed at 5:00 a.m. He hears the neighborhood dogs barking as the paperboy tosses the morning papers from his bicycle. He didn't sleep much during the night suffering distress from yesterday's late afternoon emergency leaders meeting. The lack of sleep, he reasons, has slowed the night and postponed the looming crisis.

Walking down stairs, he pauses at the front door. The paper thumps on the front step. He's already aware of the contrived journalism about him and his company. The reporter who tipped him off was clear and concise. What he doesn't know, and it's at the heart of his anxiety, is how his company's business investors will react.

Kyle Kranzton is an inventor and self-made businessman. After graduating with honors from Massachusetts Institute of Technology, Kyle set his ambitions on the desalination industry. He developed an efficient system that transforms water into steam at temperatures considerably below the boiling point. Kyle's desalination process costs about half of a typical steam-based method. In addition, his patented condensation system efficiently removes over 75 percent of saline particles during water conversion.

To finance California Water Assurance, Inc., Kyle solicited investment funds from family, friends, and interested financiers. In the first six months, the Pacific based plant produced twice the forecasted amount of fresh water. Investors increased, the process improved, and Kyle's desalination system was being heralded as the model for steam-based saltwater conversion.

Unfortunately, the new enterprise suffered two unanticipated setbacks during those first months. A freak car accident killed the company's CFO, Kyle's close friend and business partner. In addition, Kyle signed the dismissal paperwork for a gifted female technician when her resume proved erroneous. Because of a European Union recommendation, Hans

Guttmann, a pro forma expert from Switzerland, replaced their CFO. The technician position, occupied by Teresa Bellow, is still unfilled.

Picking up the mid-March daily paper he winces at the headline on the front page: *Kranzton Under Seawater.* The full front-page article includes pictures showing the technician in an evening gown opposite Kyle wearing a tux. It displays the photos as if they're socializing together. As he reads the piece, the story follows exactly what the reporter explained. Ms. Bellow claims Kranzton fired her to cover up his inappropriate sexual advances. She's suing California Water Assurance, Inc. for one hundred million dollars in damages, for loss of contract and injury to her professional career.

After the reporter's warning, Kyle discussed the accusations with his family and company management. His wife decided she and the two children would spend time at her mother's avoiding the expected media confrontations. The company's top managers requested their legal counsel to draft a public denial, explaining the true reason Ms. Bellow was terminated.

This slander and absurdity came totally unexpected. Tom Blake, the company's attorney, however, advised that the negative publicity's just a bump in the road. He stated that the primary concern of investors is the overall financial stability. Most venture capitalists are satisfied to know their investment contributions are in escrow and never commingled with the general ledger.

Kyle flips through the paper to see if it includes any other articles about his reported affair. He finds this heading on the editorial page: *Requesting Full Investigation of CWA.* Reading thoroughly, he discovers an editor making a public request for a state-appointed auditor to investigate California Water Assurance's financial records. The article assumes that Kyle's sexual antics will cost the investors thousands of dollars. In addition, the author implies that Kyle's lack of professional discipline shows he's not a competent businessman. The editorial concludes that CWA should be closed right away until investors can elect whether to withdraw their funds.

Kyle groans as he finishes the article. He knows a state auditor will take months investigating the company's finances. During that time, if shut down, CWA's loss of revenue would bankrupt their operating capital. Although he's confident that an auditor would declare no improprieties, none of that matters if the company goes belly up during the witch hunt. He takes a long shower and dresses for the office.

Kyle's standing in the company's boardroom overlooking the Pacific. Fifteen minutes have gone by, and Tom's not returned with a revised public response. Seated around the large conference table, four trusted leaders also stare towards the ocean, engrossed in their own distraught thoughts. A few minutes ago they were all debating, arguing, and yelling. Each one was giving Tom direction on how to revise his draft of a firm denial of the affair. Now out of sheer frustration, they silently wait to hear the redrafted version.

"Mr. Kranzton," Kyle's admin, Yvonne, interrupts, "Jim Cole's on the phone from First National. He says it's urgent."

Just then Tom dashes into the room, just missing Yvonne, and falls into his seat. Kyle side steps the call. "Yvonne, tell him I'll call him later." Kyle sits at the head of the table and says, "Let's hear what you've got, Tom."

Again, Yvonne interrupts. "Mr. Kranzton, Jim says it's a priority. He must speak to you now."

"Kyle, he's probably seen the paper," Tom says as he passes out copies of his revised public statement. "Put him through. Maybe we can glean something from his reaction."

Kyle hits the blinking light and puts the phone on speaker. "Jim, you're on the board room speaker phone. The top company leaders are here except our CFO. What's on your mind?"

"Well, hello everyone," Jim says with caution. "Kyle asked me to notify him promptly if anyone adjusts the company's financial accounts. Even though I have your signed letter, Kyle, I thought I'd better call you."

"What letter?" Kyle quickly replies.

"Well, the letter your CFO handed me first thing this morning. When I arrived at the bank, Mr. Guttmann gave me your letter. He said it was the top priority."

"Jim, sorry, give me a quick minute." Kyle puts the phone on mute and calls Yvonne. "Did anyone prepare a letter with my signature for First National?"

"No sir. Not to my knowledge," she answers. Kyle reconnects to Jim.

"Jim, what's the letter about?" Kyle asks.

"It's very specific. First, to draft a bank check from escrow for one hundred thousand dollars to a Ms. Teresa Bellow. She's already taken the check and left. Second, to draft another bank check of the same amount to a Mr. Ruben White. He arrived fifteen minutes ago and picked up his check. I know nothing about Ms. Bellow, but I'm fairly sure White works for the newspaper. After that, I transferred the balance of the escrow funds to your Swiss bank account number."

Everyone in the room goes numb. Hans has moved the investor funds with no one's knowledge. The facial expressions of fear, shock, and depression reflect the obvious: Hans has robbed them. Only Tom keeps his head.

"Jim, just so I'm clear," Tom remarks, "CWA paid two individuals one hundred thousand dollars apiece from our escrow reserves. Then you transferred the balance of the reserves overseas. Do you have the Swiss bank account number?"

"Hans entered the number himself as Kyle's letter instructs. Am I wrong or is this news a surprise?" Jim says, sounding alarmed. "But I've got Kyle's signed letter in my hand."

Kyle, calming himself from his initial panic, asks, "Did Hans adjust any other accounts?"

"No. All other accounts stay unchanged. But your surprise explains why Hans had a rolling suitcase in tow. He's almost certainly on a plane by now."

"Jim, don't lose that original letter," Tom speaks with firmness. "It's evidence of criminal action. We'll call the FBI and be in touch with you. Thanks so much for your call."

Kyle pushes the phone button and looks around the room. In a few moments, Hans washed away everything he and these faithful business partners have worked to achieve. Hans raided the company's funds, and they even paid the technician and newspaper editor for their roles in this disaster.

After a few minutes of sighs and blank stares, Tom speaks to Kyle. "Do you want me to call the FBI office?"

"Tom," Kyle questions. "Weren't you on that call when we interviewed Hans?"

"Yes, and so was that Swiss minister who recommended him."

Kyle pushes the intercom button. "Yvonne, can you come in again?" She promptly enters.

"Yvonne, didn't we record the interview with Hans Guttmann?"

"Yes, it was an overseas call, and we asked permission?"

"Could you find that recording?"

"Yes, sir. Do you want it right away?" Yvonne replies with a professional tone.

"Yes, but first get that Swiss guy on the phone. I think his name was Switzer?"

"Finn Schweitzer," Yvonne replies.

"Whatever. I want him on the phone right now. And don't take no for an answer!" Kyle's voice increases in volume.

Looking at everyone in the room, Kyle calms himself again. "We need to get our investor's money back. Before we contact the FBI, let's see what this guy can tell us."

Everyone waits for the overseas connection. As they sit, they phone family members on their mobiles, muttering in low voices with looks of despair. Suddenly a phone button lights up and Yvonne runs in pointing at the phone.

"Hello," Kyle says, "I'm calling about Hans Guttmann."

"Mr. Kranzer," Schweitzer replies with an agitated voice. "I don't know you and I don't know Frans Guttmann. I only agreed to answer because your mistress is so obnoxious. You must have the wrong information, so I'm hanging up."

Kyle responds, "But we have you on . . ." Tom motions Kyle to stop. As he stops, the line goes dead.

"I don't want to remind him about our recording," Tom explains. "Let the FBI handle that." Tom looks around the room. "I think we need a new public statement."

Approximately 6,000 miles away, Minister Finn Schweitzer puts down his desk phone with an ear to ear smile. As he leans back in his chair, he looks at Karl and exclaims, "Erfolgreich! That was about Hans. I think he's finished his assignment and added over two million US dollars to our coffers!"

Chapter 8

It is Sunday morning and beautiful white lilies line the granite steps. Everywhere you look, women are wearing pretty dresses with matching hats, a few adorned by young girls and many by grandmothers. As the congregation enters the church, the ushers hand out programs, shake hands, give frequent hugs, and consistently announce "He Has Risen."

Beth promised to meet Conley and Bradley at their church. She prefers to drive by herself, timid of being seen arriving with the Trumans. There's already been many questions from fellow teachers about the time she spends with Bradley and his dad. Moving into their home for two weeks added to the school gossip. Although her fondness for Bradley continues to blossom, she's aware the connection might tarnish their reputations.

"Let's go inside," Conley urges. "I don't think you should wait for Beth at the church entrance. A lot of school personnel attend here."

Bradley reluctantly agrees and follows his dad into the sanctuary.

As he gazes around the congregation, he spots Beth talking with another Springfield High teacher. He nudges his dad, and they make their way towards her.

"Don't sit with Beth," Conley murmurs. "We'll sit in the row behind her."

Taking their seats, the teacher nods to them and finishes her conversation with Beth. Beth turns to greet them.

She whispers, "Thanks for not sitting with me. I didn't realize how many school staff attend here. I hope this isn't Mr. Maxwell's church. But don't misunderstand, I'm glad you invited me." Beth smiles and reaches out to shake Conley and Bradley's hand.

"I see a visitor," Mr. Maxwell says, coming up from behind. "This is the first I've seen Ms. Williams at our church. Welcome."

Maxwell holds out his hand and Beth gives it a firm handshake.

"How is it you've attended this morning?"

"Conley and Bradley invited me to visit today. It seemed like a proper day to attend church."

"Why do you three seem inseparable to me? It makes me very jicky, as you already know."

"Jordan, we're applying for a patent on Brad's science project," Conley responds. "Since he's a minor, Beth functions as his authorized sponsor. Mr. Galloway would have been his sponsor. I think I mentioned this before."

"Remind me again. Why, Ms. Williams?"

Beth answers. "It happens Brad's project falls into an area of my expertise. My graduate degree incorporates the very elements of climate change his project aims to investigate."

"That all seems rather convenient. How about winning the state science competition?" Mr. Maxwell smirks tongue-in-cheek to Bradley.

"I plan to win both the state and national. I'll be expecting your support," Bradley grins.

Mr. Maxwell turns away and heads back to his seat. Beth smiles at Bradley, "Nicely put," she says, giving him a wink.

The organ music ascends and the overhead video screen lights. A song leader, bearing a guitar, moves toward the center of the large platform and begins playing. Soon, the large stage fills with other musicians and a small choir. As the lyrics to "Up from the Grave He Arose" fill the colorful video screen, both choir and congregation stand and sing. Before the song's over, the audience is clapping along and enjoying the annual celebration of Christ's resurrection.

After a half dozen more victorious songs, an offering, and a few announcements, they introduce the guest speaker for the service. He's a middle-aged missionary from India, an international speaker. During his sermon he shares a good deal of humorous travel related stories. His Indian-English pronunciation keeps everyone's attention.

As he closes his message, he challenges the congregation with an unusual question. "What's the one thing in life you can't possibly live without?" To some parishioner's annoyance, he repeats the question several times for emphasis.

Finally, he answers his own question. "I'm sure there are many answers in this room to the question I propose. Let me suggest the best answer is this one: the presence of the living God. The presence of the living God should already be with you. It's the one thing that continues with you when you pass from this life. Never reject his resurrection and his resurrected presence. God bless you."

Bradley feels those words penetrate his heart. For a moment he thinks about his hopes for a science career. He wonders about his future with Beth. As he stands to sing the closing hymn, he questions his faith. He thinks about graduating into real life in three months. He's singing a song, but thinking it's time to follow God in a more serious way. *It's time for more faith,* he thinks, but he wonders how that will come about.

The congregation stands after the pastor pronounces the benediction. Conley and Bradley linger to talk with several of their church buddies. Beth makes her way to an exit. Once out in the parking lot, she's welcomed again by a few fellow teachers and invited to return. She climbs in her Geo and checks her phone. She waits for Conley to text her the address of a restaurant. When she receives it, she types back "OK, see you there."

Arriving first at the restaurant, Beth sits at the reserved Truman table. She's emotionally exhausted. She wonders if this cloak and dagger connection to Bradley makes any sense. It's hard for her to be honest with her feelings when her reputation's constantly on the line.

When Conley and Bradley arrive at the table, she emotes frustration.

"I think Mr. Maxwell was standing on the church steps when I left. I felt I was being watched. The church service was great, but I can't go back. It seemed like every teacher was expecting me to do something stupid." She stops and breathes deeply and reaches into her purse for a pill. Neither Conley nor Bradley respond.

Conley picks up his menu as if to ignore her comments, but Bradley takes Beth's hand. "Are you okay? I'm sorry this morning stressed you out."

"I'm fine, really. Just needed to get that out. What are you having for dinner?" she smiles and grabs her menu.

A server arrives and takes their drink orders. The three debate their best meal selection. Beth relaxes, and Bradley drops his concerned look.

While waiting for their meal, Conley reaches into his suit coat pocket and pulls out an envelope. "I received this late yesterday, but waited until now to show it to you." As the letter passes from Beth to Bradley, big smiles come on their faces.

"Looks like we just had another resurrection," Bradley exclaims. "How long does this provisional patent last?"

"For the next twelve months we can claim patent pending status. However, we must complete the patent within that time. There are no extensions. You two need to shake down the system, so we can file the final paperwork."

"Does this mean we can display desalination at the state science fair this Friday?" Beth asks.

"If you're ready to exhibit desalination, go for it." Conley smiles. Beth and Bradley give each other a thumbs up.

Conley puts the letter back in his jacket. "This document goes into our safe deposit box and a copy in our home files."

Friday morning, the three pack the F-150 with the desalination equipment and drive to Dayton. The Ohio State Science Fair occurs each year during Easter break, and although Bradley's visited the fair previously, today he's eager to be a presenter. Beth rides up front with Conley, and Bradley sits in the back, reviewing his notes and diagrams.

"To increase the desalination percentage, it seems to be a balance between the microwave and the wire loops. Can you think of any other variable I'm missing?" Bradley utters.

"Give it a rest champ," Conley laughs.

After a few minutes of silence, Beth responds, "What about the thickness of the nickel plating? I don't remember ever challenging that spec."

"Good point. I got the thickness from something I read about water stretching. I've always just used the same thickness."

Conley joins the conversation. "What about water speed? If we slow the water, would it have more time to work?" No one responds.

"Hey," he chuckles, "I'm not a scientist."

"No, we're thinking," Beth exclaims. "That's a good suggestion."

"Actually, they're both good thoughts," Bradley states. "They represent two areas our tests haven't included. We need more tests before the national."

Bradley and Beth discuss how to test both ideas and before long Conley's driving down South Jefferson Street looking for East 5th Street.

"It's just ahead, Dad," Bradley says, seeing a street science fair banner. Conley turns left into a long line of traffic. Contestants are waiting to unload their exhibits at the Dayton Convention Center entrance.

Bradley's worried about the delay. "It's 10:30 and we need to be ready by 1:00. I'll hop out and see if there's another unloading area."

While Conley and Beth wait in the pickup, Conley questions Beth. "I hope you don't mind me asking, but do you feel the same way about Bradley as he does for you?"

"Brad's so wonderful to me. The more I spend time with him, the more I care for him. I guess I would answer 'yes' to your question, but we seldom talk about our feelings." She pauses. "The constant scrutiny of our relationship is so frustrating. It's probably the same for him. I'll be glad when he turns eighteen and graduates."

Bradley steps up to his dad's window. "See that black traffic attendant with the orange hat. The one waving at us. Follow him to a loading dock. I'll meet you there."

Conley pulls out of line and follows the traffic attendant. "I want you to know, if you become my daughter-in-law," he says, "I'll be very pleased. However, I just don't want you and Brad to get sidetracked by a stupid mistake. It's only another two months. After that you can spend time together anywhere you want."

"Thanks for looking out for us."

The attendant waves Conley towards the number four loading dock. As he backs in and parks, Bradley arrives to help unload. In fifteen minutes, they assemble the IronWorks' frame and roll it into the hall.

The large convention hall is decorated with school banners from all over the state. As they advance to the Springfield High colors, Beth hurries ahead to greet her other students setting up their exhibits. Linda comes to meet Bradley.

"When did you leave this morning?" she asks.

"Around 8:00," he responds. "Not sure exactly when we left. We picked up Ms. Williams."

"So she rode with you the whole trip? That must have been cozy."

"Ms. Williams rode in the front seat with my dad. I sat in the back reviewing test plans." He smiles and lightly fist bumps her on the shoulder.

"Why are you looking at test plans? After today, your project's old news."

"Yeah, what about nationals?" Bradley says as they position the frame under the Springfield banners. "I might need to fine tune this puppy again to take top honors."

"I hear your 'BS' pomp and circumstance," she laughs. Beth motions to Linda and asks to review her electrical exhibit.

Bradley tests his power connections, boots up his computer, and sets the wire loop in the channel. As he's testing the reservoir water tank for saline, he gets a tap on the shoulder.

"Oh, hello, Mr. Maxwell."

"Brad, now how's that patent coming?" Mr. Maxwell says sarcastically.

"I brought a copy of our provisional patent approval." Bradley digs through his backpack and hands him the paper. Mr. Maxwell reads the document and hands it back.

"Your dad mentioned this at church last Sunday, but I didn't think he was serious," he responds somewhat humiliated. "Good luck today." Mr. Maxwell makes the rounds to the other Springfield High students.

Each of Beth's students is prepared for examination by the over seventy-five volunteer judges, comprising high school and community college science teachers. The competition's scheduled from 1:00 to 6:00. The exhibit judges review student projects for about twenty school districts. With a little over six hundred school districts in the state, the judges overlap districts to provide for impartial scoring.

Part way through the afternoon, Mr. Galloway comes to the Springfield High exhibits.

"Hey, remember me," he smiles. "How are you getting along with Ms. Williams?"

"She's a lot prettier," Terry smirks.

"I'm sure," he grins. "Springfield High's on my judging schedule, so let me take time to review your exhibits."

He walks around noting names and projects on his evaluation paperwork. He deliberately goes to Bradley's exhibit last.

"Did you meet your water stretching goal?" he says to Bradley.

"Today's exhibit is on desalination, converting seawater to fresh water. You can review the recorded saline parts per million on the white board each hour," Bradley says proudly.

Mr. Galloway's quiet. He carefully looks at Bradley's updated computer illustrations. Watching the water flow and saline foam, he puts his finger into the foam and tastes it. He makes several notes, reviews the ppm reports, and walks away without another word.

At five o'clock a slick Ohio State congressman takes the stage microphone. "The judges have completed their initial evaluations. Everyone's done an excellent job again this year. We're all very proud of our science minded high school seniors. We'll be announcing the winners in thirty minutes, so presenters are welcome to look around the hall."

Bradley shuts off the water flow, but leaves his computer program running. He starts to stroll around the hall, but hears Mr. Galloway call his name.

"Bradley, can you stay with your exhibit? I want to show a few people."

In ten minutes, thirty to forty people, each with a judge's badge, surround the IronWorks' frame. Mr. Galloway introduces Bradley as a former student and asks Bradley for a desalination demonstration. Bradley briefly rechecks his equipment and turns on the water. As he turns to the crowd, he sees Beth among the spectators.

"I need a helper to record my findings. Ms. Williams, my Springfield High science teacher, will you please help me?" Beth comes forward and steps to the whiteboard.

As Bradley takes water readings, Beth marks the ppms on the board. The low reservoir tank has a higher saline concentration than normal, so the ppm readings look exceptional to the judges. After the presentation has been completed, Bradley and Beth receive applause.

"Just one question before we disperse," Mr. Galloway shouts. "Have you applied for a patent?"

Bradley grins. "We received a provisional patent one week ago." Again the crowd applauses.

As the judges make their way backstage, Bradley and Beth shut down the water and electric to pack up the equipment. Linda comes over and takes Bradley by the arm.

"Impressive, I admit," she whispers. "You may go to nationals after all. Can you take me with you?"

"Sorry, but I'm guessing my dad will want to be my sidekick at nationals."

"Will Ms. Williams go with you?"

"That's up to her. She's my patent sponsor, but I don't think she's required to be there."

Linda looks at Bradley with a sober look. "Why do I think she's my competition?"

"I don't know, Linda. I've never been on a date with either of you."

The congressman returns to the microphone. "We've decided on fifteen third place prizes in five categories, eight second place prizes in four categories, two first place prizes for two categories, and one overall Ohio state science fair winner." The exhibitor's excitement grows as the speaker lifts the cloth cover off several tall, shiny trophies.

With no pause the congressman announces third place prizes. The categories include engineering, electricity, communications, human/animal health, and nutrition. Among the winners, Bradley's classmate, Karen, gets a third place trophy for her flu virus deviation exhibit.

After awarding fifteen happy third place winners, the speaker gets right to the second place winners. Eight proud science students receive second place trophies for their hard work.

"We have two first place winners; one for agriculture and one for climate change. The winner of the agriculture category is Kathleen Joseph from Cincinnati for her exhibit on hybrid seeds. The winner for climate change is Bradley Truman from Springfield for his demonstration of converting seawater into fresh water." Both Kathleen and Bradley come forward to receive glittering trophies.

After the applause, the congressman allows a few seconds for a strategic pause and awards the final trophy. "This year's Ohio state overall science award goes to a well-deserved student. Someone who's not only shown an important facet of climate change, but has taken initiative to apply for a patent on his work, Bradley Truman. Bradley, please join me on the stage."

Bradley walks to the stage with a standing ovation of the judges and participants. He looks confident and stands straight and tall. Shaking hands with the congressman, he receives a four-foot-tall silver and gold trophy. He notices Mr. Maxwell, who's returned for the award ceremony, and motions for him to come on stage. Mr. Maxwell comes beaming and shakes hands with Bradley and the congressman.

A long line of attendees waits to congratulate Bradley. He gets dozens of questions about patent applications. When he gets free, he realizes Beth and Conley have packed everything and loaded it into the F-150 truck bed.

Meeting them at the loading dock, he's elated and exhausted. "This was incredible," he says to his dad as he carefully loads his awards and hops in the passenger seat. "I've never felt so good about anything. We're going to nationals." Looking at Beth, stretched out in the back seat, he whispers, "How's she doing?"

"She's already half asleep. It's time to head home."

As the three travel back to Dayton, a trolling motor pushes a black inflatable almost silently through the Persian Gulf. A new moon shines somewhere in the skies, but April rain clouds hide both the moon and any star light. It's pitch black. The boat with three passengers is invisible from the Ras Al Khair shoreline. Two of the passengers, wearing diving gear with bulky oxygen tanks, quietly slip backwards into the sea. They carry a plastic panel with metal reinforcement that unfolds for their purposes. The boat operator turns in a wide circle and motors back to their secret place of origin.

The seawater intake pipe is a twenty-minute swim underwater. Reaching the intake pipe, the panel's quickly unfolded and precisely positioned to stop further intake of ocean water. The suction within the pipe holds the panel firmly in place. The divers return to the rendezvous location. As the canvas boat returns, the operator helps remove the divers' tanks and gear before they board. Once the diving paraphilia's loaded into the boat a handgun's grabbed. The operator pumps both divers in the head and they bob in the Gulf. Trolling back to the hidden dock, the boat operator dials a nearby cell number on his mobile. When the line connects, he speaks one word "Erfolgreich," which means "successful," and tosses the cell into the sea.

A plant engineer's been waiting the last two hours in the shadows of the Ras Al Khair desalination plant. After receiving the call, he pockets his phone and grabs his tool kit. Using his assigned keys, he accesses the security gates, entrances, and interior doors. He hastens down the metal stairs to the steam chambers. With the intake blocked by the underwater panel, the steam pressure's already reached above normal. As the pressure grows, he places four explosive charges in strategic areas around the chambers. He hopes the steam tank explosion hides the extra damage from the charges. The timing must be perfect, but he's confident he knows just what to do.

He sets his watch while observing the raising pressure gauges. After three minutes of careful examination, he calculates the steam explosion moment. He'll set off the PE-4 explosive charges seconds before the steam tanks explode. Someone will eventually discover the charges, but the exploding hot water and steam tank wreckage will cover the evidence for weeks.

He hurries out of the plant and returns to his vehicle. With his eyes set on his timepiece, and his window open, he can hear the metal strain and screech from the growing heat and pressure. At the precise moment, he triggers the charges and watches the facility explode. He grabs his phone and dials a Bern number. When the line connects he says, "Erfolgreich", tosses his phone out the car window, and speeds away.

Chapter 9

THE FIRST WEEKEND IN May, Springfield High School takes a Thursday and Friday teacher break. Other schools across the nation follow a similar schedule allowing teachers to prepare for the final weeks of the school year. This is the time for high school seniors to compete for national science awards hosted at the National High School Senior Science Rally, commonly called the NHS3R. The location for the Saturday competition is the Bartle Hall Convention Center in Kansas City, Missouri.

The drive to Kansas City's a little over six hundred miles. Conley rents an enclosed U-Haul trailer to carry the exhibit framework. Traveling together gives Bradley and his dad time to chat again about the loss of a wife and mother. They consider Bradley's future, his science career, and discuss his relationship with Beth. Nights include Bible reading and prayer, as well as ice cream and Netflix.

The Trumans arrive at Bartle Hall mid-afternoon on Friday. After receiving directions from a convention center security guard, they unload their trailer at one of the double door entrances. A NHS3R assistant aids them in locating their assigned area, and the convention center staff guarantees the safety and security of the exhibits. Working together, they assemble the display. Bradley salts the holding tank water, tests the power connections, and checks the charge on the twelve-volt battery. By 6:00 p.m. everything's prepared, so they wander around the huge hall watching as dozens of other exhibits take shape.

"Wow! Must be a hundred exhibits," Bradley exclaims.

"Well, at the least one from each state," Conley replies. "Do you see anything like ours?"

"I can't tell. Most of these displays are a mystery. I hope we get a chance tomorrow to check them out." Bradley looks at his watch. "Beth should be here soon. Let's go to the hotel to meet her."

Conley promptly reserved rooms for the three of them after winning the state. He chose a Kansas City hotel within walking distance of the convention center. As they walk with their luggage to the hotel, the city lights revitalize their tired bodies.

Pushing into the lobby, Bradley spots Beth in line at the hotel reservation desk.

"You're here," Bradley says, surprising her. Beth steps around her luggage and gives Bradley a hug.

"Our flight landed early. Is everything ready?"

"If nothing changes before 10:00 a.m. tomorrow, it'll be ready, set, go," Bradley replies. His excitement regarding the competition is obvious.

"Have you eaten?" Conley asks her.

"Not really. Since lunch I've just had airplane peanuts and airport coffee."

"Should we eat in the dining room or order room service?" Conley asks.

Bradley and Beth answer him in unison. "Room service."

The hotel lobby's luxurious. As Conley registers them, Bradley can't help but notice the shine on the marble floors. Mirrors cover the area's interior pillars, and much of the spacious area has brown leather couches and chairs in cozy arrangements. The lobby leads right into a dining room and hotel waiters bring fancy drinks and light desserts to guests in the leather seats. What he doesn't notice is two older European gentlemen. They are carefully observing the three of them register and disappear into an elevator.

In the morning, the trio proceeds to the hotel dining room for breakfast. The dining room ceiling goes to the top floor of the hotel. The walls display huge photos of Kansas City's prominent sites. Bradley and Beth grab cups of coffee, find a table, and admire the city scenes. While Conley scans the breakfast bar, they get a chance for private conversation.

"I wanted to be with you last night after we ate," Bradley confesses. "I thought you might want to gab a bit."

"It's okay, I showered and went right to bed," she replies. "I was zonkered. When do we leave for the convention center?"

"The contestants can arrive after 9:00, but the center doesn't open to the public until 10:00." Bradley gets up to fill his plate. Beth follows as he heads for the breakfast bar.

As Conley starts back to their table, his mobile rings. "Hello?"

"It's Mark," a familiar voice says. "I heard you won first prize at the state and assume you're going to the NHS3R, the National High School Senior Science Rally?"

"We're in Kansas City now," Conley replies, finding his seat at their table.

"I wish you'd called me. I appreciate your weekly technical reports, but you're not keeping me up-to-date with your schedule. You're trekking into dangerous territory. People at the NHS3R will offer to purchase your patent, some for profit and others to bury your technology. Remember, you've only secured a provisional patent. You can sell your research, but not a provisional patent. I'd strongly recommend you resist any offer until we can vet the buyer and your patent's approved. Please keep me in the loop."

"So you're saying we should expect proposals?" Conley questions.

"Yes. Like I said, some will be legitimate and others will be scams," Mark responds. "Desalination's an economic and political minefield."

"Thanks for the heads up. I'll improve at keeping in touch." Conley hangs up as Bradley and Beth join him with their plates filled with breakfast tidbits. As they eat, Conley relays Mark's message.

Entering the convention center, he and Beth start through the exhibit checklist. However, they're unsure if they've enough saltwater for an eight-hour water flow. The fifty-gallon tank's full, but they decide to only turn on water when busy. When the water channel's not flowing, they'll answer questions and use the computer program. The exception, they agree, must be when the judges are present.

By 10:30 a.m. a reasonable number of browsers have strolled into the hall, so Bradley starts the water for the first desalination ppm posting. After Beth takes the readings, the first numbers on their white board are impressive: water tank ppms, 33,437; post nickel plating ppms, 33,256; foaming water ppms, 687 ppms; and before collection tank, 682 ppms. For the national competition, they brought a bucket to catch the foam, which often drips to the floor. On the side of the bucket, Beth painted "Brine for Sale."

A few college science majors introduce themselves. Their curiosity is only in the water stretching at the bridge. Beth provides satisfactory answers and they continue their stroll around the exhibits. It's almost noon before anyone stops by who understands desalination. Bradley's alone while Beth's talking with a college rep and Conley's away purchasing lunch.

Two men, examining the white board, approach Bradley. One man has gray hair and a European cap that sets just above a silver earpiece. He puts out his palm for a handshake. The other man, slightly taller with a collarless suit jacket, also shakes hands with Bradley.

"Good day," the graying man says with a hint of German, "My name's Fredrick Weber. My associate and I are very interested in your display. May I ask, do you think you can compete with current desalination technologies?"

"Mr. Weber, thank you for your interest," Bradley responds professionally. "I believe the most common process being used today is a multi-stage flash distillation. However, with our process, although the water's heated, it does not require steam to separate the saline particles.

"Comparing our process to the popular reverse osmosis system, we use less hardware. There's no membrane to deal with or mechanism required to operate it. If we agree the primary need for new technology is to reduce costs, we believe our process keeps costs lower."

Bradley pauses for a moment before continuing. "But, to answer your question, there's a lot of new desalination technology in the making. It's too soon to tell how our method will compare to others."

"That's a smart answer," Fredrick responds. "Your project has interested us for a few months. I hope you're not offended, but we photographed your high school exhibit and the contest in Dayton." The taller associate pulls a set of prints out of his inside coat pocket and hands them to Bradley.

Bradley looks at the prints and the flattery results in a huge smile. Seeing Bradley's positive facial response, Fredrick continues the conversation.

"We think your technology may be useful in Europe. We've many well-trained scientists who can build on your foundation. This process could become a first class desalination plant for Spain or Italy. How does that sound?"

"Well, that sounds great," Bradley grins. "I'd love to go to Europe."

Glancing at his associate, Fredrick responds, "Talk with your team. We'll talk with our superiors. Perhaps we can arrange an agreement beneficial to all." With that comment, the two turn and walk towards an exit.

Conley arrives back with deli sandwiches and bottled water. "Where's Beth?" he asks.

"She's talking with a few college guys. She must have wandered off with them. I met two men from Europe." Bradley grabs a sandwich. "Dad, those European guys want us to work with their scientists in Italy or someplace like that. Wouldn't that be awesome?"

"You didn't say 'yes'?" Conley questions.

"Well, no, but I want to," Bradley says with excitement. "They'll be back later to discuss it."

Beth arrives talking about her acquaintances. She reaches for a sandwich and a bottle of water. "I saw you talking with those two German looking guys," she says to Bradley. "How did that go?"

"Great, they invited us to Europe."

"Remember what Mark told us this morning," Conley says in a low voice. "Not all these offers are legitimate. We must be cautious."

Just then five well-dressed individuals appear and each one's wearing a brightly colored judge's badge. Beth and Bradley put down their sandwiches and start the water flow. For ten minutes, four judges inspect everything: the water channel, the bridge, the twelve-volt loop, and even the nickel coating. The fifth judge studies the computer program as it loops over and over. Beth posts a new set of numbers and the judges ask a few questions. As quickly as they appeared, the judges disappear moving to another exhibit.

Before the three can huddle to assess the judges' evaluation, two cowboys in Stetsons, jeans, and western boots swagger over to inspect the exhibit. One, with a mustache, attentively checks Beth's white board numbers, pulls a tablet from his backpack, and does several calculations. The other, exposing his shaven head, closely examines the water channel from beginning to the end. Seeing the foam dripping into the bucket, he snatches a bit with his index finger and licks it. A smile comes on his face.

"My name's Matt Timmons," he says after licking his finger. He holds out the same hand for a handshake and nods towards his companion. "This is my brother Paul." Paul puts his tablet away and stretches out his hand too.

The three introduce themselves.

"I never make a deal without a taste test," Matt laughs. "Which one of you tasted the foam?"

Beth grins. "I did."

"Now you must be Bradley's little sister. So tell me why you tasted the foam?"

"I wanted to taste the sodium residue to test the fresh water conversion," Beth replies with a red face. She prefers not to address the little sister comment.

"You are one smart little girl," Matt grins. Bradley's offended by Matt's comments and opens his mouth to defend Beth, but she motions for him to hold his tongue.

"Did you know the ocean water has millions of molecule-sized sediments? Most of it is sodium chloride. In desalination plants, sodium chloride is a waste product and dumped back in the ocean. To us, however, it's a gold mine." Matt nods to Paul.

"Sodium chloride's big in the medical industry," Paul begins. "Its use is becoming unlimited. Every day medical research discovers new ways to use it for good health. That's our business, the business of Aquatic Mining Company. At AMC, we provide refined sodium chloride primarily for the medical industry. We convert unused foam, the "brine," into a useful form for research and production. We've been in business for ten years and business is booming. To be honest though, desalination plants are changing their policies, and it's getting harder to procure good brine . . ."

"So we," Matt injects, "plan to invest in our own desalination plant, produce our own brine. In our case, fresh water'll be the by-product. Cool, right?"

"The point is," Paul continues, "we'd like to proto-type your system to see if it'll work for us. Come to San Diego this summer and let us test your system in a real environment."

"That's quite an offer," Conley replies, "but we'll need time to consider it."

"I understand," Paul retorts. "You need to talk to your people. We need to talk with our people. Then our lawyers review contracts, consider patent implications, negotiate expenses, blah, blah, etc. But once it's all agreed, bah-dah-bang, bah-dah-boom, you're in lovely San Diego."

"I understand testing the system in a larger environment, but we've already proven we're successful at desalination," Bradley says.

"Yes, and no," Matt answers. "You have yet to prove you can meet California regulations," he winks at Conley and continues, "and your method's cost effective. There are two other desalination exhibits here. They're both based on thermal methods, and we already know saltwater conversion by steam is too expensive. That's why we're making this offer to you."

"Well, we're definitely interested," Conley replies.

"Great, let us take you to dinner tonight. Our lawyer's John Timmons, another brother. We'll call him and see what he suggests. How about we meet here at 6:30? Will that give you time to pack?"

Conley looks at Bradley, who nods his head. "Thanks for your interest. We'll see you tonight."

As Matt and Paul saunter off, Bradley's over the top. No one can wipe the grin off his face. "We've had two good offers, one in Europe and one in San Diego. If there's only two other desalination exhibits, we've got a good chance at high marks in the scoring."

"I need a break. I'm off to the washroom," Beth declares. She turns to Bradley. "Are you okay by yourself for about ten minutes?"

"Sure. Pick me up another bottle of water."

Conley grabs his mobile. "I'm giving Conti a call, but I'll be nearby."

The crowd lightens after the lunch hour, so Bradley shuts down the water. Conley, on his mobile, steps to the side to allow curious attendees to approach their exhibit. It's quiet for the next several minutes for father and son, but not for Beth.

As she exits the washroom, she looks for a water bottle machine. Without delay, a well-dressed woman in her fifties hastens to her. "Are you Elizabeth Williams?" the woman, with a French accent, questions.

"Yes," she responds, "but please call me Beth."

"So you're with the Truman exhibit?"

"Yes."

"I read your master's degree paper on climate change. It was . . . well, . . . wide-ranging, but confusing and full of false assumptions. It's hard to believe you've any understanding of climate change. In fact, because of your reputation, your association with the Trumans will undoubtedly destroy their chance for success."

Beth's mouth drops and her stomach churns. She's so taken by surprise she can't think of anything to say.

"Let me make you an offer," the French woman continues. "I'll make a place for you at the Eastern Hemisphere Climate Change Symposium. You can publically defend your scientific notions surrounded by people who really know what climate change is all about."

"Who are you?" Beth asks with an indignant ring in her voice.

"I work for the European Union. They're aware of your scholarship, but have taken a dim view of your published work. I'm serious, though, about including you on a desalination panel during the EHCCS. You can present your position in their backyard. Are you up for the challenge?"

"Definitely!" Beth firmly replies. "Send me the invitation and materials. I'm shocked by what you're telling me. I want the chance to prove myself!"

Beth forgets all about Bradley's water and marches back to the hall. She can't wait to discuss her unhappy meeting with an insulting French lady. As she arrives, the two German men are bargaining with Bradley and Conley.

"So after talking with our European advisors, we can offer two hundred and fifty thousand US dollars for your method. That means you'll surrender ownership and any other rights. We have committed our proposal to paper for your review."

Conley takes the proposal, steps aside, and quickly reviews it.

"Bradley," Mr. Weber says with a sales pitch, "I think our offer should cover all your exhibit expenses and deliver a satisfying profit. You'll be graduating soon and this project will hold you back from other interests. Besides, I'm guessing you can use extra college money."

Bradley makes no response, but Beth steps up to Mr. Weber.

"My name's Elizabeth Williams," she says, "are you familiar with a scholarly paper called *Inevitable Climate Change*?"

The man steps back and removes his cap, revealing a patch of black hair. He pulls his associate aside. The two turn their backs and argue in German for a short time. Then Mr. Weber slowly walks back to Bradley and Conley, looking embarrassed.

"Mr. Truman," he says to Conley. "I'm sorry, but we must rescind our proposal. We were unaware that your female associate's Elizabeth Williams. We cannot be part of this project. This would make us the laugh of the EU science community."

No one knows what to say. Beth's furious. She's insulted again by a European. She turns her back. They shake hands with Conley and Bradley and walk away. With an angry voice, Beth turns and says, "No one's going to insult me like that again. I'm going to Europe and teach them what I know about climate change!" She digs through her purse and begins chewing on an antacid.

Mr. Weber and his German friend hear her loud comment. Weber reaches into his pocket and pulls out a small flip phone. Dialing the preprogrammed number, he listens for someone to pick up. Upon hearing breath on the other end, he says, "Erfolgreich," and closes the connection. Yanking out the phone's battery and SIM card, he twists the phone until it separates. He throws the pieces into separate trash containers, and the two Europeans smile as they exit the conference center.

Chapter 10

THREE WEEKS AFTER KANSAS City, Bradley and Beth sit patiently in an airport terminal waiting to board an international flight to Brussels. During the three-week period, they've rarely seen each other outside of school. Beth's determined to bolster her knowledge of climate change and defend her white paper's suppositions. Bradley's stored away the IronWorks' frame and equipment, and he's occupied with year-end studies. Conley's commuting to Germantown working with Mark on examining the Aquatic Mining Company offer.

As Bradley and Beth walk the jetway to the business class section of the plane, it's 10:30 p.m. They're both tired from the day's preparation and excitement. Beth received two tickets from the European Union and solid confirmation she'll sit on a discussion panel during the desalination session. Everything she knows about climate change is stored on her laptop.

Both were informed they'll each have personal FBI escorts when they land. Mark Conti insisted on these protection arrangements. He's aware of the international political friction regarding saltwater conversion. The government briefed Mark's team of lawyers on the growing connections between the Nhlabane, California, and Ras Al Khair plant disasters.

Sitting together as the plane lifts off, Bradley takes Beth's hand. "Time for a quick prayer," he says. "Lord Jesus, protect us, bring us home safe, and may we hold our own at this conference. Amen."

"Thanks," Beth responds and keeps hold of his hand. "Tell me again what your dad said about AMC's offer."

"Since our dinner in Kansas City, their lawyer, John, suggested AMC license our process instead of purchasing it. Dad agrees it's a better arrangement. We hold all the ownership rights and they pay for the right to use it, the same as renting a piece of property. The discussion price is one hundred fifty thousand a year, but we need to prove we can meet the requirements."

"So can we license other companies at the same time?"

"I guess," Bradley answers. "The lawyers can figure that out."

"If it can work out, I'll feel a lot safer in California. I'm nervous about going who-knows-where across the pond," she sighs. "Right now I'm tired. I'll nod off any minute."

They lean their seats back, both turn off their overhead seat lights and get comfortable. Beth closes her eyes, but Bradley's too excited. He brings up a game on his phone to help him get to sleep.

Three hours into the flight, Bradley wakes to Beth's sniffles. He gently takes her hand again. "Are you okay?" he asks.

"I'm sorry. So sorry I've let my pride separate us the last few weeks. I've been so angry and it's just my own egotism," Beth whispers and wipes her tears.

"Where's this coming from?" he probes.

"I woke, opened my Bible app, and read about Jesus' humility in Philippians. Conviction consumed me. I've been so wrong about this trip."

"It's okay. We got a free ride to Belgium."

"It's not okay. I need to trust Jesus for my reputation, not the European Union. I've got to adjust my attitude before the meetings. I need your help. Please."

Bradley pulls her head to his shoulder. There's a lot he wants to say, but he answers, "I'll do whatever it takes." He takes her hand and they pray together.

In the morning as they make their way through the Brussels airport, the terminal's flooded with arriving international passengers. They progress slowly through passport security and find their way to the baggage claim. As they lean on each other waiting for their luggage, two individuals in dark suits parade around the baggage conveyor displaying their names on small placards.

"Over here," Bradley yells. He motions to a black female who arrives with a white male companion.

"Good morning," the petite woman says, putting her hand out to Beth. "I'm FBI Agent Corina Reynolds. I am assigned to you while you're in Brussels."

"I'm FBI, Tom Phillips," the man says, motioning to shake Bradley's hand. "I'll be your shadow the next two days."

At that moment the buzzer on the baggage claim blasts, the carousel moves, and everyone steps up to reclaim their belongings. It takes about twenty minutes for Bradley and Beth to locate their suitcases. Agent Phillips signals for a taxi.

"I thought the EU would shuttle us," Beth says.

"We notified the EU office we would arrange travel to the hotel," Agent Phillips states as he sits up front with the cab driver. "Take us to the Groen Internationales Hotel."

"No problem, Tom," the cab driver replies in a Brooklyn accent. The driver laughs as he looks in the mirror at the surprised reactions from Beth and Bradley.

Agent Reynolds smiles. "He's with us."

As they arrive at the hotel, the cab driver stops in front of the doorman. The driver assists Beth and Bradley with their luggage and murmurs to Agent Phillips. "Let me know if you need me again." Phillips nods.

"Don't you guys have any baggage?" Bradley asks.

"We arrived at 7:00 a.m. and moved in," Agent Reynolds remarks. "We examined the suites and swept for bugs," she smirks, "both kinds. The hotel has accommodated all EU panelists on the fourth floor. The security's increased for those suites and throughout the hotel premises."

"Do we need to check in?" Beth asks.

"Yes, the rooms are in your names. You're the official guests."

As Beth and Bradley wait in line, the agents keep watch around the reservation area. Noticing the long line of others checking in, Beth and Bradley enjoy foreign fashions and overhear unknown languages. Every so often they poke each other and nod in a certain direction to bring attention to some strange outfit or hairdo. Despite the lengthy wait, they complete registration and catch the elevator to the fourth floor.

Beth's suite is near the elevator at the center of the long fourth floor hallway. As she opens the door, she's delighted by the large suite. To the right of the door is a small kitchen with a nearby dinette table which overlooks the main street. The rest of the large space has Queen Anne styled furniture with flowery upholstery. It matches the draperies and blends with the carpeting.

"This room's so lovely, it reminds me of my bedroom before Dad and I moved out of our home," she says to Reynolds. "Where are the bedrooms?"

Agent Reynolds steps to the left wall and swings open the hidden bedroom door. "Walk forward to the windows and you see the doors. I've put my stuff in the bedroom by the kitchen, so you can have this one. I hope you brought a plug converter for your hair dryer."

Reynolds opens the bedroom door and Beth walks in with a huge smile. "I love these European room themes; the colors are so vibrant. Does your bedroom have the same colors?"

"Yes, the whole apartment follows an English theme, but it's not the same as Bradley and Phillips' suite. Theirs is a French Country theme. I suspect each suite on this floor displays a different country design." Reynolds pauses and asks, "What did you mean about leaving your home with your dad?"

"My mom moved out right after my twelfth birthday. I'm not sure what it was about, but both Mom and Dad assured me it was not because of me. Anyway, he sold our home, and we moved into an apartment near his work. I miss my old bedroom. I've been in dorm rooms and apartments since sixteen."

"Have you talked with your mom since she left?"

"No. It's like she vanished. The same's true for my dad. The week after I started college, he disappeared. I've not seen either one since. But someone paid all my college bills, including an allowance. I've always believed it was my parents."

"So you've been on your own since sixteen? That must have been hard, but you have a handsome gentleman now."

"Bradley's great, but it won't last. Everyone I love leaves." Beth sighs as she lifts her suitcase on the bed. She spends the next several minutes unpacking. When finished, she tests her computer charging adapter on the hotel wall plug.

"Everything's good on my computer," she says to Reynolds.

"Okay. Let's go visit with Bradley and Phillips."

Meanwhile, Bradley's getting settled a few doors down the hall. While he unpacks, Phillips starts twenty questions.

"If you don't mind me asking, what college do you attend?"

"I'm a senior at Springfield High School in Springfield, Ohio."

"So what's Ms. Williams to you? You seem close."

"She's my high school science teacher." Bradley keeps on emptying his suitcase.

"I thought you two were together. It surprised me when they said you'd be rooming separately."

"We spend a lot of time together because of our science interests, but we aren't in an actual relationship."

"So they invited her to speak, and she brought you along for the experience, or what?" the agent probes.

Bradley stops and faces Phillips. "What's your concern?"

Agent Phillips takes a more serious posture. "The FBI has asked me to risk my life for a friend of an EHCCS panelist, but I don't understand why. You're not a family member, you're not her assistant, you're not her lover, so who are you?"

"I'm the person who invented a new desalination technology. Beth's the patent sponsor and has the credentials to represent me. We're a team."

"Okay. You're the cake and she's the icing. I can live with that." There's a knock on the door and Agent Phillips answers. Beth and Agent Reynolds march in and tour the men's suite.

"You're right," Beth says to Reynolds. "This room is a different theme and color scheme. It's a little feminine for these boys don't you think?" She laughs as she puts her arm around Bradley. "Just kidding."

For the next hour, Phillips and Reynolds go over various security protocols for the hotel and the conference center. Not knowing what to expect, they offer alternative scenarios if an unexpected hostility should arise. Phillips gives each a business card for the FBI taxi cab driver in case they get stranded or lost.

"It's our intention to stay close to you. If that's not possible and you need help, call the number on that card," Agent Reynolds stresses.

"Now that we've got you both scared by all these security rules," Agent Phillips declares, "we believe people will be friendly. We suspect the conference will bore all of us, and you'll be safely on your flight home tomorrow with no troubles at all."

"How about questions?" Phillips asks.

"What's the plan for the conference?" Bradley replies.

Agent Phillips and Reynolds explain how the day's activities will follow the EU agenda.

At 11:00 a.m., the four ride the elevator to the hotel lobby. Each dress in their conference attire: Beth in a light green pants suit with a matching colorful blouse; Bradley in a beige sports jacket with dark pants and cream-colored shirt, no tie; and both agents . . . like FBI agents. All are wearing EU name badges provided by the EHCCS representative at the hotel.

Along with several others wearing badges, they're escorted into a large hotel ballroom. The byzantine tapestry and granite sculptures surrounding the room catch Beth's eye. Bradley gapes in awe at the huge crystal chandelier in the middle of the ceiling, lit with dozens of candles. At one end of the room is a long banquet table filled with cuisine from several nations.

Beth and Bradley make a start for the table, but an English scientist she had met at Stanford distracts her. Soon she's involved in multiple introductions and talking with other EU panelists from the various panel sessions.

Bradley nods to Agent Phillips and the two of them grab a plate. Making their way down the banquet spread, they read the calligraphy notes describing the various foods. Without making a detailed examination, they agree there must be dishes from every EU country. Bradley's too hungry to wait for Beth, so he and Phillips sit at an empty table and begin eating.

"Excuse me," a woman says. "Excuse me, are you a scientist?" At first, Bradley doesn't realize the young Italian woman's speaking to him. She taps his shoulder and asks again.

"Why, yes," he says. "Although I'm not as educated as most of these participants."

"Few of us are. You look very young," she responds. "I'm in my second year of graduate school. It's a little overwhelming." She pauses and asks, "So how did you get an invitation?"

"I'm here with Ms. Elizabeth Williams. I'm working on an invention and she's my sponsor. She invited me to attend."

Ignoring Agent Phillips, she continues speaking with Bradley. "I think I know of Ms. Williams. Did she graduate from Stanford and write a paper about climate change?"

"Yes."

"I'd love to meet her. All across Europe, we study her paper. I'd guess everyone in this room has read it either by choice or requirement."

"I heard the European community disliked her climate change ideas," Bradley replies.

"Not so at all. She's very well respected. I begged for an invitation just to hear her speak on the panel. Everyone was so pleased when she agreed to come."

"It was nice talking with you. You're welcome to come over when Ms. Williams joins us. I'll introduce you."

Bradley turns his head back to the table. He ponders the negative comments about Beth from the two German men in Kansas City. His mind's deep in thought when Beth and Agent Reynolds sit down and join them.

As Beth starts to eat, the Italian woman returns. Bradley taps Beth on the arm and says, "I promised this lovely lady I would introduce you."

Beth turns and shakes the woman's hand as she says, "Your Stanford paper was wonderful. It helped prepare me for the science of climate change. Thank you for shaking my hand. Ciao."

As the woman walks away, Beth looks around the table and speaks quietly. "Everyone I've talked to has read my Stanford paper. Someone told me it's required reading in most European colleges for beginning science. We've been lied to. My own pride has gotten us here."

Looking directly at Agent Reynolds, she whispers, "It's very alarming."

At 12:30 a ringing bell signals the large group to board two chartered buses. Agent Phillips jumps the line, heads to the back of the second bus, and stands alert watching each one board. Beth and Bradley follow Phillips, and Reynolds instructs them to sit together in a middle section. As Reynolds follows them on the bus, she walks slowly to prevent the passengers from crowding to close. Both agents are attentive to anyone who gets near enough to harm either of them.

A few passengers joke as if this is a sightseeing excursion, but the bus driver's announcement is clear. "This bus is for EHCCS panelists and their guests. We'll be going directly to the EU headquarters where they'll provide your conference room assignments. We'll drive you back to the same hotel at 16:30 this afternoon. If you finish early, you're welcome to wait in the bus."

The buses travel the narrow back streets of Brussels. The ride takes twenty minutes and passengers are rather quiet. Many panelists are visiting Brussels for the first time. They peer out the windows at the city activities. When the buses arrive, the passengers exit at a rear loading dock of the EU complex. The panelists enter the tall dock doors where the EU staff identifies their designated conference room locations.

Beth and Bradley, with their FBI escorts, follow the guide's instructions to the third floor and enter room number three hundred five. It's a large open space filled with chairs. Phillips guesses the room seats about two hundred and fifty. At one end is a small stage with seating and microphones for a facilitator and three panelists. Behind the stage, and all along the eastern wall, are floor to ceiling windows with spectacular views of

Brussels. As they walk towards the windows, a well-rounded female brings Beth a panelist's name badge.

"Ms. Williams, I'm the group monitor for this session. It's such an honor to have you on our panel," she says with a Danish accent. "We're so pleased you accepted our invitation."

"Thank you. I'm pleased to be here." Beth smiles, accepting the ornate panelist badge.

"In a few minutes the guests will enter," the woman continues, "and we'll start at 13:30. The discussion should be over between 15:30 and 16:00. Oh, I see the other two panelists have arrived, so let's meet them." As the monitor moves away, Beth looks at Bradley and shrugs her shoulders.

Agents Phillips and Reynolds take strategic positions in the room. Bradley sits near the front to provide Beth moral support. As Beth sits at one of the panel seats, dozens of guests move into the room. At 13:30 the female monitor stands, taps her mike for a sound check, and requests everyone's attention.

"Thank you for attending this first Eastern Hemisphere Climate Change Symposium. Today we are discussing 'Fresh Water in a Changing World'. Let me introduce our panelists and we'll begin."

Beth's stomach does a flip. She's nervous about how the audience will respond to her. She reaches into her purse for a chewable tablet. After introducing each panelist, the audience provides a friendly round of applause. As the monitor introduces Beth, the audience gives her a standing ovation. Beth's astonished, but Bradley's proud and gives her a wink. She pulls the mike close for a grateful, "Thank you so much."

The plump monitor opens the floor for discussion and the subject of finding fresh water during progressing climate change begins. Representatives from various EU countries speak regarding their country's unlimited fresh water resources. Non-EU country delegates from Iceland and Norway boast of their fresh water statistics, snow and ice conversion methods, and renewable supply. Company executives from Evian, Smartwater, and Icelandic Glacial, among others, brag about their production and distribution of bottled water worldwide.

The two panelists, when questioned, respond positively to the discussion. However, each panelist cautions that bottled drinking water may become inadequate should climate change predictions come about. One panelist, taking exception to one water bottling executive's statistics, quotes the UN prediction that approximately 14 percent of the world's population will encounter water scarcity by 2025.

At that remark the monitor questions the audience, "How will we provide fresh water for those people?" There's a unified response from water bottling company representatives. They declare their production can meet any new requirements through proper distribution. The representatives claim sufficient fresh drinking water reserves can solve any water shortages.

The monitor turns to Ms. Williams. "I'd like to hear your thoughts on this issue."

Beth picks up the microphone and stands. "I thought I'd stand or no one would see me," she laughs. The audience laughs along with her and it lightens the room's growing tenseness.

"This is an Eastern Hemisphere conference, not a Northern Hemisphere conference," she states. "This conference includes the Middle East and Africa, and certain other arid regions of this hemisphere.

"I know fresh drinking water is available across Europe and North America through natural reserves. However, over the last several years, over twenty thousand desalination plants have converted seawater in over one hundred countries. Many plants are in the Middle East and Africa. Several are operating in Europe, the United Kingdom, and the United States. The fact is this planet is already depending on seawater conversion to meet needs for irrigation, city water requirements, and certain industry necessities as well as drinking water.

"Every person in this room understands the basic truth about the earth's water supply: 96 percent of the earth's water is seawater, with saline levels too high for human consumption. The total earth's potable water supply is the remaining 4 percent. About 70 percent of that fresh water's unavailable because it's locked up in ice caps and glaciers. There's no doubt that tapping into seawater is necessary for future survival."

As Beth takes her seat, the room explodes into squabbles. Delegations from the Middle East and Africa dispute the bottling company's claim about sufficient worldwide resources. Faces become red with anger and the crowd noise increases with pointing fingers. The tension in the room grows as does the aggressive body language. The monitor bangs on the table, but no one pays any attention. After a few minutes, she calls security.

Within thirty seconds twelve heavily armed EU security officers ring the outside of the room. The audience quiets. The monitor asks for the audience to choose representatives to speak on their behalf. After fifteen minutes of noisy chatter, four individuals step forward.

"One at a time please," the monitor demands, pointing first to a tall black African woman.

The woman is handed a microphone. "Please, we in Southern Africa need desalination and new desalination technology. Purchasing bottled water from Europe is very expensive and distribution isn't consistent. We've yet to find a distributor willing to provide enough water for our farm animals and cooking needs. We don't understand how European water will be available to meet those needs."

The monitor looks at the three remaining group representatives. "Which of you can provide the water resources to meet Africa's needs?" she asks.

A well-dressed European steps forward. "Brine's the key problem with desalination. Only a part of the saltwater's converted to fresh water. The remaining unconverted water is returned to the ocean with a higher percentage of salt content. Besides the high content of saline in the brine . . ."

"Excuse me," the monitor interrupts. "You're not proposing a solution, but just pointing out the deficiencies within the desalination process. Can you provide us with a solution to Africa's water scarcity?"

The man drops his head for a minute and then replies, "The bottled water production in Europe comes from an unlimited supply of fresh water. This supply, in any required quantity, can be shipped anywhere in the world."

"Thank you," the monitor says. "Now who will be next?"

A man dressed in beige robes and a red turban steps forward. "In Saudi Arabia, we desperately need water for our city's population. Partially for irrigation, but mostly for the large city's substructure. I'm not addressing European drinking water. We import bottled water for drinking, but to date no bottled water company has provided any practical means of distribution to meet our non-drinking water requirements. If they did, we would weigh the cost of European supply against the expense of investing in desalination plants."

As the Arab steps back, the fourth group representative steps forward. She waits for the monitor's approval to speak. When the monitor nods, she begins.

"I'm a nurse and Peace Corps worker from Australia working in Kenya. I'm assigned to a large hospital complex in Nairobi. Having been in

Kenya for almost five years, I know firsthand about the increasing droughts and difficulties providing sufficient water for agriculture and animal life. However, as I listen to the predictions of a rising ocean, I'm worried about our small supply of fresh water being contaminated with saltwater. If the predictions are true, we'll not only see a decrease of our African seacoast, but an increase of brackish groundwater. I'm not aware of any European water company addressing these needs." Again the room erupts with noisy chatter.

The monitor stands again, waiting for the audience to calm down. "I know many representatives from bottled water companies are in attendance. Can any of you provide insight into your company's plan to address these nondrinking water requirements?"

For the next few minutes, it's very quiet. No one comes forward to answer. The monitor looks at the panelists and says, "It's 15:45. I believe we've learned something important today about the need for both pure drinking water and the desalination process for agriculture and city infrastructure. Would each of the panelists please give your thoughts before we adjourn?"

After the first two panelists speak, Beth offers her opinion, "There's no doubt climate change, in some form or another, is happening. Let me admit that the world's science community is only guessing at the outcomes and the timing. However, we've established in this room that regions with ample fresh water supply are not prepared to distribute sufficient water to areas with arid climates. If such distribution was already in progress, we wouldn't have so many countries using various desalination technologies. The challenge is to find economical ways to supply fresh water for every region's needs using the technology we have, and can develop in the next twenty years."

The monitor thanks everyone for their attendance and closes the discussion. As people exit the room, the uproar between Europe's fresh water companies and desalination supporters continues. Beth takes Bradley's hand and they hurry to the bus with the FBI agents on their heels.

Chapter 11

THE CHARTERED BUSES LEAVE the European Union headquarters and soon return to the Groen Internationales Hotel. Bradley remembers the bus trip to the EU was rather quiet. However, the ride back is noisy with bottled water promotors arguing vehemently with those who champion desalination technology.

Because of radio and TV broadcasting of the conference discussions, reporters, photographers, and curious spectators crowd the buses' front and rear exits. Exiting their bus and pushing past the cameras and inquiries, Beth, Bradley, and their two FBI companions escape into the hotel lobby.

"I'm exhausted," Beth complains. "I need to lie down before tonight's session. Glad we're flying home tomorrow."

"After we relax a bit, I'll order room service," Bradley suggests. "I'm more hungry than tired, but I'll check with you before I order anything."

Beth and Agent Reynolds take the elevator. As they enter their suite, their hotel phone rings. Reynolds answers it and hands it to Beth.

"Hello, Beth Williams here," she says, trying not to sound worn out.

"So pleased I caught you," says a woman's voice. "I'm Alessia Amato, President of Eau Suprême. I attended your panel discussion today, and I wanted to say thank you for your involvement. I won't keep you. I'm an old friend of your dad's, so tell him hello when you see him. Ciao." Beth doesn't remember any female VIP and doesn't want to hear about her dad. She lies on her bed and falls asleep.

Before going to their suite, Bradley and Agent Phillips meander into the hotel dining room and examine a copy of the room service menu. With help from the concierge, they get the menu translated into dishes that make sense for Americans. Satisfied they can help Beth and Reynolds order room service, they ride to the fourth floor.

Bradley tosses his suit jacket on the sofa and lies down. After Agent Phillips checks in with his superior, he sits up and engages the agent in dialog.

"I'd say our panel discussion got quite heated," Bradley remarks. "What do you think, Agent Phillips?"

"I'm not paid to think," the FBI escort teases with a grin.

"That's not the first time you've said that," Bradley laughs. "But what was your opinion of the room dynamics?"

"Bottled water's a big industry in Europe. It's what keeps thousands of people working. The EU's not ready to share its water business with a competing method of providing the world's water."

"It reminds me of the Y2K dilemma," Bradley says thoughtfully.

"You weren't even born yet," Phillips chuckles.

"Yeah, I know, but the Y2K challenge was similar. First, whether to even believe a technological disaster was imminent, and second, to agree on how to solve the problem before it occurred. In those days there were several vested interests which made any Y2K solution a stand-off. Sounds like the same thing."

The agent chooses not to respond. Bradley picks up the menu and reviews the meals with Phillips.

"Agent Phillips, I've decided what to order for dinner. Have you given it any thought?"

"Absolutely, I'm going for the carbonade flamande. It's the one meal I've craved since I learned we'd be in Belgium. It's a national dish."

"You know it's made with beer and you're supposed to drink beer with it. But you're on duty protecting my life." Bradley chuckles as he walks towards the suite door.

"Well, no one worries about the beer in the stew," Phillips quips. "And I'll have a bottle of EU water from our suite refrig. If you're going to Beth's room, let me check the hallway."

Phillips looks both ways and then motions Bradley into the hallway. Bradley softly raps twice on Beth's door. Using a door knock code to aid in security, Agent Reynolds raps back two knocks and Bradley raps again with five knocks.

Agent Reynolds sticks her head out of the door. She peers up and down the hallway and motions him inside.

"Beth's sleeping," she says softly.

"No problem," Bradley responds. "She needs her rest. She usually only naps about thirty to forty minutes."

While Bradley makes small talk with Agent Reynolds, a man at the hotel bar orders a German beer and slowly sips. The bartender can't help but notice his crooked nose, which holds up thick eyeglasses. The tapster also observes the customer's mustache is a black and gray color, which is different from a thatch of blond hair emerging from under his German cap. After a few sips, a mobile rings. The beer drinker pulls the phone from his inside coat pocket and listens. One word's spoken, "Gehen."

The man pockets his phone. He makes a discrete nod to an attentive hotel bellhop, who rapidly disappears down a stairway. The drinker pays for his beer and walks out of the nearest hotel bar exit. Once on the street he removes his glasses and mustache tossing them into two different nearby trash containers.

Moments later, the fourth floor elevator doors open and eight hotel stewards emerge with a bottle of champagne in one hand and an ice bucket in the other. Although the fourth floor houses fifteen suites, the champagne's only intended for guests in eight rooms, including Bradley and Beth's suites.

The eight stewards simultaneously knock on their assigned room's doors. All eight recite the same greeting in broken English. "It pleases this hotel you are a guest. We offer this decanter of champagne without charge."

Agent Phillips steps to the door and gazes through the keyhole. He recognizes the hotel's uniforms, but delays his decision to open the door. In the seconds he waits, the steward pulls a 9mm handgun from the ice bucket and fires five rounds randomly through the door. The first two strike Phillips' Kevlar chest protector and propel him backwards. The third one goes through his left arm, and the final two bullets sink deep into his pelvic area, one bullet penetrating his femoral artery. Phillips drops to the floor, blood streaming from the last two wounds. With three quick kicks, the shooter boots the door open, ignores Phillips, and straightaway rushes to find and assassinate Bradley. When Bradley's not located, he pulls out his com and reports, "Truman missing." Following protocol, he hurries back into the hallway and rushes down a stairwell to escape capture.

At the instant the aggressor shoots Phillips, assailants attack seven other guest rooms. Agent Reynolds goes to her suite door hearing the steward's greeting. Not wanting to disturb Beth, she replies, "Thank you, but we're not interested." After her remark, the steward drops his ice bucket, and she hears the metal bucket bounce against the door. In one quick motion she jumps aside and signals Bradley to hit the floor. Five bullets fly through the door and into the sofa where she and Bradley were talking.

As the shooter kicks at the door, Reynolds grabs a metal kitchen bar stool to brace against the door latch. She motions Bradley to Beth's bedroom and shakes her head "no" when he starts to close the bedroom door. She runs to the apartment size refrigerator and pushes it from the wall sufficiently to hide. In moments the door crashes open.

Meanwhile, Bradley finds Beth trembling on the bed. Her face is wet with tears and she's holding her mouth closed so not to scream. Bradley whispers to her and lifts her in his arms. He carries her to a large wardrobe diagonally positioned in the bedroom. With an adrenaline tug he moves the wardrobe enough to expose a small, but sufficient, hiding place. He tenderly sits her on the floor and puts his arms around her. They sit motionless, praying the aggressor, now in the suite, will not find them.

When the shooter breaks through Beth's door, he rushes into the suite and heads to the bedroom with closed doors. As he dashes through, Reynolds emerges from her hiding place and shoots him in the buttocks, intending to injure and disarm him. The aggressor quickly turns his gun on her, so she pumps three shots into his chest.

As the man falls, Reynolds bolts into Beth's bedroom calling her name. Beth and Bradley crawl out from behind the wardrobe.

"Are you okay?" Reynolds asks.

Bradley nods, but Beth can't speak. She's still trembling. Reynolds directs Bradley to place her on the bed and props pillows under her feet. "Take care of Beth. I want to check on Phillips," she exclaims.

She pulls out her com and requests emergency backup as she heads down the hall. She dodges guests running through the hallway for the stairs. Rushing into Phillips' suite she finds him on a bloody floor, his pulse's gone. From the amount of blood loss, she doubts medical support could have saved him. Peering into a few other rooms with dead bodies, she concludes Beth and Bradley may be the only targeted guests not assassinated.

Rushing back to Beth's suite, she can hear the whirl of a helicopter on the hotel roof. Three US Marines in combat uniform charge up the hallway and run into Beth's suite. Agent Reynolds barks orders.

"Put this woman on a stretcher. She's in shock and needs oxygen. The man," pointing to Bradley, "and I will follow you to the roof."

As the Marines secure Beth, Reynolds calls out to Bradley, "Where are the documents, computers, stuff we need to take?"

"Beth's laptop contains all the desalination documents. There's nothing in my suite that matters."

Grabbing Beth's laptop off her bedroom dresser, Reynolds hurries toward the roof stairway with Bradley on her heels.

"Where's Agent Phillips?" Bradley yells to Reynolds.

Reynolds shakes her head and keeps on running. Bradley instantly understands, but decides he'll let his grief and questions wait until the chaos subsides.

The two rush across the roof to the waiting chopper. Reynolds makes sure Beth's comfortable and Bradley's safely on board.

"I'll see you both later," she yells and turns back towards the hotel stairway. The blades churn and the chopper lifts off. Bradley can see the streets below. They're jammed with ambulances, fire trucks, and police cars.

The EU security team and local Belgium police are in gridlock as they argue over jurisdiction. The murder of international guests makes for a complicated circumstance involving multi-national police, bodyguards, and Interpol. No one's sure who's in charge or how to investigate the deaths on the fourth floor. Only a few even notice the US Marine helicopter lift off the roof and disappear across the city.

One man, in a German cap and dark trench coat, waits near a streetlight for the Belgium police to get the crowd under control. He watches the US helicopter land and depart.

An EU delegate, commanding attention at the hotel entrance, summarizes the situation. He speaks in broken English using an electronic bullhorn.

"Everyone, please, please, calm down. I'm from the European Union offices. Let me state that the EU security team is in charge of this investigation. The Belgium police force will support their efforts. All ambulances, fire trucks, and other emergency vehicles should only take direction from an EU security team member.

"Concerning today's tragedy, a group of assassins masquerading as hotel employees invaded eight suites on the fourth floor. They murdered guests in six suites. In two other guest rooms a bodyguard and an assassin were killed. The guests for those two suites are currently missing . . ."

The man wearing the trench coat is disgruntled. He strolls away from the crowded area and opens his flip mobile. When the line opens, he reluctantly says "Erfolglos," meaning "unsuccessful." He breaks the phone into two pieces, and as he walks away, kicks the phone pieces into the street sewer drains.

Forty minutes later, the helicopter hovers over a NATO controlled airport. Beth's strapped and connected to an oxygen mask. As the chopper begins its descent, Beth lifts up her head and motions to Bradley. He releases her mask.

"Am I riding in a helicopter?" she asks excitedly. Bradley nods.

"Holy kamoley," she retorts. Then leaning back with a smile, Bradley reconnects the oxygen.

When the propeller blades slow, a Marine speaks to Beth. "Are you able to walk on your own?"

"I think so," she replies. He unstraps her and disconnects the oxygen apparatus. The Marine assists her and Bradley to the ground. Bradley clutches her as six combat ready Marines surround and march them towards a US airplane about thirty yards away. Stopping at the entrance ramp of the modified Boeing 757, they're met by a Marine in parade dress uniform.

"I'm Marine Corpsman Dr. Dennis Jacobs. How are you both doing?"

"I'm better, but freezing. I can't get warm," Beth replies.

"After what you've been through, I'm not surprised. Can you climb the boarding ramp?" Beth and Bradley nod. Dr. Jacobs follows them up the ramp. Reaching the aircraft entrance, the corpsman seats them in a large open area just beyond the plane's plug door. The seating area contains two beige leather sofas with four matching chairs arranged in a conversation pit fashion.

"Give me a few minutes to prepare everything," Dr. Jacobs says as he hurries off to an inside cabin.

Beth and Bradley try to relax in the plush furniture. It's only a moment before Beth smells the aroma of freshly brewed coffee. She stands and looks around the room. There's a writing desk in one corner and a large storage cabinet near the conversation pit. Next to the pilot's entrance, she spots a side room and heads for the opening. She comes back with a large china cup of steaming java.

"There were no seats this comfortable on the flight coming to Brussels," she whispers as she sips the hot coffee.

When Dr. Jacobs returns, Beth asks, "What kind of airplane is this?"

"I cannot answer that question," Dr. Jacobs laughs. "It's above my pay grade."

He takes Beth by the arm. "I've arranged a hot shower for you in the executive washroom. When you're finished, you can put on one of the blue robes until your clothes arrive from the hotel."

"Thanks so much. It sounds delightful." She hands her coffee cup to Bradley.

Dr. Jacobs hands Bradley a handwritten airplane menu. "When I come back, be ready to order for Beth and yourself. Food prep takes about twenty minutes. By that time, Beth will be back to eat with you."

When Dr. Jacobs reappears he relays Bradley's dinner selections on the intercom secured above the writing desk. Bradley listens to the Marine's shorthand version of beef stew and mixed vegetables, but turns to the aircraft's entrance hearing a familiar voice. Agent Reynolds sees him and exhales with relief. She motions to the Marines to deliver Beth's and Bradley's personal articles to Bradley.

"Beth's okay?" she asks without delay.

Bradley rises and embraces Reynolds. "Yes. You saved us both," he says warmly. "I'm so sorry about Agent Phillips."

She acknowledges his remorse with a slight nod, but says, "I picked up your clothes and personal items. I hope I didn't miss anything."

"Weren't you afraid to go back to the hotel?"

"Getting your personal effects wasn't the only reason I went." She takes Bradley's hand and beams proudly. "You were very brave today under extremely difficult circumstances. Your cool head was a significant help to Beth and me."

She turns, pulls out her mobile, and slips down the plane's hallway. Connecting to her superior, she confirms Beth and Bradley are safe. She reports clean fingerprints were collected from the dead assailant and verifies she's stripped Phillips of all identification.

Beth returns in a large blue robe as a steward carts in their meal. Seeing her suitcase, she prefers to get dressed before eating, but, smelling the beef stew, she decides she's too hungry to wait. With the steward's instructions, both pull meal trays out of the arms of the furniture.

A Marine officer enters, salutes, and announces, "We'll be taking off in two hours. These seats are yours for the eight-hour trip to Cleveland. We've arranged for Mr. Truman to meet us at the airport. You're not permitted in any other section of the plane due to security clearance constraints."

Pointing as he speaks, "You can put your belongings in this cabinet. A washroom is behind the cabin door. If you need anything, you're welcome to use the intercom. We have assigned a Marine to check on you during the flight."

At that moment, the officer stops speaking and comes to attention. Entering the plane is a tall gray-haired politician recognized all across the US, Vice President Kennedy Kleaver. Beth and Bradley promptly push their trays back and stand up in respect and amazement.

"Sit down, sit down, please," the VP says cordially. "Welcome to Air Force 2. At least that's the term the FAA uses. You kids had quite a day. I'll bet you didn't expect this much excitement when you arrived in Brussels this morning."

The VP makes himself at home on a couch. "Now finish up your dinner. While you're eating, I'll give you the background information. It's classified, so you can't repeat this to anyone."

Beth's so surprised at meeting the vice president she can't finish her meal. Bradley seizes the opportunity to finish his beef stew and Beth's vegetables. Vice President Kleaver begins his explanation.

"There's been a growing animosity between the water bottling companies in Europe and those places in the world using desalination. It's really all about money. It's fair to say Europe has much of the world's best fresh water and thousands of Europeans make a living supporting one aspect of the water business or another.

"With the growing concerns about the rising oceans, the European Union allowed the fresh water companies to unite under the name Eau Suprême. You may have heard that name at the panel session." Bradley and Beth nod.

"Eau Suprême's really a political arm for the European Union fresh water industries. The organization is challenged to protect Europe from technologies that undermine their fresh water business. Due to climate change predictions, this has become increasingly complex both politically and practically. As you're aware, in the last few years, thousands of Syrians flooded into Europe bringing all kinds of political turmoil, not to mention, crime, unrest, and prejudice. The prospect of a mass immigration or military invasion from regions suffering from the lack of water has caused Europeans to be fearful. The same could happen again, only worse."

The VP stops as the steward enters the room and quickly cleans up Beth and Bradley's dinner trays.

"Are you following me thus far?" Beth and Bradley nod affirmatively.

"Okay, now this is the classified stuff," he says, lowering his voice for emphasis. "We have reason to believe a militant force within Eau Suprême's organization desires to crush desalination technology. This means the destruction of effective desalination plants, the eradication of new saltwater conversion technology, and the elimination of desalination advocates.

"Last winter, someone mysteriously incapacitated a new desalination technology being tested in South Africa. This spring, a large desalination project in California went belly up due to the embezzlement of their investors funds. Only eight of the fifteen suites on your hotel floor were attacked. In each case, they were panelists at today's conference favoring desalination."

Bradley speaks, "Do they have any idea who's organizing these attacks?"

Vice President Kleaver stands, "We think we have a clue, but still working on it. Perhaps the fingerprints on the assailant in your hotel suite will help us."

"I'm telling you this because you've invented a new desalination technology. They have targeted both of you. Therefore, we are assigning you security at your home and school. You're welcome to go about your daily routine, but you'll have FBI surveillance twenty-four by seven. You're not restricted from discussing your experiences, but use discretion. And no mention of Eau Suprême or US suspicions."

Another full dress Marine enters the room, and the VP follows him. As he leaves, the VP turns and says, "I'll not see you when we land. Perhaps another time. God be with you both. You're very courageous."

As Vice President Kleaver disappears into the plane's inner hallways, Beth and Bradley look at each other with astonishment.

"What have we gotten into?" Beth whispers.

"The question is, 'What has God gotten us into?" Bradley responds.

As the plane taxis, they store their suitcases and personal items. Sitting next to each other, Bradley takes Beth's hand, and she leans her weary, wet head against his shoulder. After the plane levels off, Beth enters the washroom, dresses, and dries her hair. Before she falls asleep with her head on his chest, Bradley gives her a little poke.

"Holy kamoley?" he asks teasingly.

"It's an expression my mother used to say," she responds as her eyes close. Exhausted, they sleep through the flight. When awakened by the steward, they're descending into the Cleveland Hopkins International Airport.

When the high security plane lands it does not taxi to a public terminal. An FBI Suburban meets the VP's military staff on the landing strip. Bradley, Beth, and Agent Reynolds hop into the FBI van. A Marine packs their luggage in the back, and they drive to a high security airport terminal. The Boeing 757 takes off again within a few minutes.

Conley waits outside his F-150.

"Are you guys okay?" he asks as the three exit the FBI vehicle. He immediately hugs Bradley and Beth and says, "Agent Reynolds, I'm so glad you were there." Then he gives Reynolds a hug too for her heroism.

They put their luggage in the truck bed and Agent Reynolds gets into the front with Conley. As Bradley and Beth settle in the back, Conley hands an envelope back to Bradley. When he sees the return address, he lets out a big "Yahoo!"

"What's that?" Beth teases grabbing at the letter.

"It's my freedom ticket," he wisecracks. "My first step to independence. My wings of flight." He grins at her inquisitive face and says calmly, "My senior driver's license."

"When did you turn eighteen?" Beth asks. Everyone can tell she's annoyed.

"Last week before we left. We were so busy getting ready for our trip I hardly noticed. We'll party in a couple of weeks with my second step . . . graduation."

Conley yells back to Beth. "The FBI's suggesting you move back in for a couple weeks. Any problem with that?"

"It'll help us with our protective surveillance," Agent Reynolds adds.

"That's fine. I'm part of the family now," Beth laughs. "I just need a few things from home."

"Well, please don't go to your apartment," Reynolds cautions. "If you need anything, I'll send someone to get it. Starting tonight you'll be staying at the Trumans. For the next few weeks, we'll provide twenty-four by seven surveillance for the three of you. We've already contacted Mr. Maxwell."

"Does anyone at school know about what happened?" Bradley questions.

"I don't know for sure," Conley replies, "but the incident made the national evening news. Somebody at school must know about it. Tomorrow will tell." Turning to Reynolds, "What can anyone say?"

"The classified information from the VP is off limits," Reynolds states firmly. "Don't mention the VP's plane or anything discussed with him. You can answer questions about anything else, but try to avoid the press."

"Are you spending the night too?" Conley asks Agent Reynolds.

"I'll be picked up at your home, and another agent will be on all-night stakeout. Since these two," pointing to the back seat, "slept during the whole flight, I'll miss all the fun at your place tonight."

Chapter 12

STEERING HER CAMRY INTO the USPTO office parking lot, FBI Agent Missy Kearn notices a black Ford panel van in a parking space. She's concerned at once. This secluded office facility and parking area is specifically off limits to the public. She slows to check out the suspicious vehicle. The van displays no license plates, no window registration, or any other identifying marks. As she waits in her car, she reports the vehicle to the local FBI office. After receiving instructions, she exits the Camry. Drawing her weapon, she inspects the van's exterior, attempting to unlatch a door. All van doors are locked tight. Agent Kearn replaces her gun in her shoulder sling and reports her findings. The FBI promises to send a towing service.

It's now 8:25, and she doesn't like to be late. Since being assigned this duty, it's Missy's practice to always open the office window blinds, turn on the HVAC, and start a pot of coffee before 8:00 a.m. She routinely switches the night answering machine to the PBX console and transcribes any overnight messages. Serving as the receptionist and admin for the three resident patent attorneys takes up most of her day, but it's not her primary responsibility. If there is a security threat, it's her task to guard the classified patent information stored in an enormous vault on the ground floor.

As the pot brews fresh coffee, two of the patent lawyers arrive. Making their way past Missy's desk and through the customer workroom and lounge, they greet her and ask about the van. She explains a tow truck will arrive soon and hands them copies of a few overnight messages. After they enter their private offices, she unlocks the vault just as Mark Conti, senior attorney, arrives.

"Good morning, Missy," Mark smiles. "It's already sunny and warm outside. What's the story on that panel van?"

"The van's locked. I couldn't open any door. Anyway, someone's coming to tow it soon. I'll let everyone know when it's gone. But while it's here, I'm designating the office security level at yellow."

"Gotcha," Mark grins. "Any messages?"

"I'll bring you a cup of brew when I finish your messages. Then we can go over today's schedule."

Nodding, Mark heads to his office.

As she hears Mark close his office door, she sees a flatbed tow truck pull into the parking lot. Three men exit the towing vehicle and begin walking around the van. When Missy looks up again, she notices the tow truck trio has opened the back doors of the van.

"Why haven't they backed the tow truck into loading position?" she mumbles to herself. Staring at the tow truck, the workmen emerge from the back of the van dressed in long coats. They run towards the USPTO entrance.

In one well-trained motion, Missy presses the silent alarm under her desk, changes the interior security to level red, and electronically locks the front entrance. As the men arrive, she dons her headset and waits to speak on the entrance intercom.

Each lawyer sees the status change, but two of them are deep in internet research and delay their response. Mark's waiting for his cup of coffee and catching up on various overdue filing. Seeing red, he clears his desk, locks certain files in his cabinet, and grabs his Kevlar vest. He ducks into the bathroom to put on the protection.

When the long coats reach the front door, they discover it's locked.

Missy addresses them over the intercom. "Do you require assistance?"

All three respond by pulling automatic weapons from under their coats and shooting into the glass doors. In moments they've shattered the security glass and kicked their way inside. They tramp to her desk, yelling and waving their weapons.

A tall assailant yanks all the wires from the PBX phone console. Missy jumps off her chair, but he grabs her shouting an ill-mannered, "Sit down, shut up." For the next ten seconds, they empty their clips at everything in the lounge: tables, desks, computers, cabinets, and wall hangings. They demand the attorneys reveal themselves.

As the armed men pace back and forth, two attorneys enter the bullet-riddled customer area. With the assailants watching the lawyers step into sight, Missy takes the chance to sneak away. She steps into the vault and closes the door. She's trained to protect the patent documentation from the inside. The vault's equipped with an outside phone line; a loaded Glock, with extra clips; Kevlar protection; and a secret release button which opens the vault door when the threat has cleared.

The man in charge is a short stocky bearded man with a French accent. He points his AK-47 at Joshua, who is the youngest of the lawyers. "I want all information for the Truman patent. I want history, diagrams, prototypes, patent application, all your lawyer notes. I want it now!"

Joshua's terrified. Stuttering he responds, "Well . . . it's . . ."

"I want to hear YES!" the man yells, jams in a replacement clip, and fires three bullets into the lawyer's chest. The young attorney drops to the floor bleeding profusely. As the gun swings to the second attorney, Mark enters the room from the bathroom and faces the bearded man.

"The answer is 'yes,'" Mark quickly states. As the shooter looks him over, Mark peers into the front office to make sure Agent Kearn has disappeared.

"Are you the last?" the man growls.

Mark nods. "Yes, there's only three of us.

"Give me all the Truman documents!" The man points his weapon at Mark's chest.

"Everything you want is in our security vault around the corner," Mark replies pointing towards the front entrance.

"Take me now and open it!"

"Opening the vault takes a little time." As Mark responds, he tries to conceal a slight grin creeping up on his face. He's facing the parking lot and sees a SWAT team positioning themselves near the office entrance.

Noticing the tiny grin and sensing trickery, the bearded man yells, "Wrong answer." He fires three shots into Mark's chest, slamming Mark into the wall where he slumps to the floor. Before a fourth shot's fired, a SWAT bullet penetrates the bearded man's back, comes out his chest, and lodges in the wall. He drops to the floor, partially falling on Mark.

The other two attackers lower their weapons and raise their hands. The SWAT commander confiscates their guns and checks them for other weapons. Both attackers slip a tiny white tablet into their mouth and collapse within seconds.

Less than four hours later, Conley answers his ringing mobile. "Conley, are you okay?" says a distressed voice.

"Mark, is that you?"

"Armed men attacked our offices today. They killed Joshua and shot me three times in the chest."

"What are you talking about? Someone terrorized your office?" Conley speaks faster and louder. "Where are you? Do you have help? Should I call someone?"

"Conley, wait, wait, let me explain," Mark insists. "This morning three men shot up our office. They wanted your patent documentation. They shot Joshua point blank and put three bullets in my Kevlar vest."

"So you're okay?"

"Yes. The only person harmed was Joshua. Although Missy complained about being in the vault for two hours." Mark pauses, but continues when Conley doesn't respond. "The purpose of my call is to warn you. Someone's after your patent information and they mean business, very nasty business. You need law enforcement help."

"Oh, I didn't call you . . . again," Conley confesses. "Last week, somebody ambushed Bradley and Beth in a Brussels' hotel. The US Marines rescued them. We're all under FBI surveillance."

Mark's quiet for a moment. "I saw that news report. What's going on? This climate change debate's turning water into a new target for profit and political gain. What'll happen to your AMC plans?"

Conley leans back in his chair. "We're not changing course. Bradley graduates in two weeks. He and Beth have already made travel arrangements for San Diego. We're going ahead with AMC just like we discussed. They've guaranteed full security."

"What about you?" Mark asks. "Who's watching your back?"

"I've installed a new home security system, and, as I mentioned, the FBI's providing twenty-four by seven surveillance. It's a good thing my office is at home, I don't plan to go out much," Conley laughs.

"You still have a sense of humor," Mark responds. "While our offices are being repaired, I'll be at meetings in DC. I'll let you know about any changes."

"Thanks for the call. I'm so glad you're all right. The Lord's looking out for you." As Conley disconnects, he hears Bradley and Beth open the front door. Coming down the stairs, he sees two FBI agents with them.

"Agent Reynolds," Conley grins. "It's nice to see you again."

"You'll see a lot more of me," Agent Reynolds says, extending her hand. "Because of the attack on the USPTO, we're assigned escort services again. Beth might as well return to her apartment. From now on, I'll be with her."

Introducing her partner, she says, "Meet Mateo Rodriquez. Matt will be Brad's teammate."

Turning to Conley, she continues, "When you leave the house, whoever's assigned to house surveillance will accompany you."

Bradley and Matt shake hands, but Bradley wonders what will happen to Matt. He pictures the last time he saw Agent Phillips alive.

The Springfield High School auditorium's packed. It's a typical hot afternoon graduation, and the school's under-capacity air conditioning can't keep the room cool. Mr. Maxell's running around trying to make sure everything's on schedule. So far the clerk has misplaced a box of diplomas, the sound system's blown a fuse, and not all the school board members have arrived.

Bradley's waiting in line with two hundred twenty-three other seniors. Under his graduation gown, he's wearing a new suit, with a tie Beth picked out, and sweat's dripping off his face. Near him stands Agent Matt Rodriquez in unmistakable FBI apparel. The last two weeks, the seniors have gotten accustomed to seeing Bradley and his constant companion. After Bradley and Beth's escape from Brussels became common knowledge, the teachers and students accepted the otherwise awkward escorts. Although a day of significance for Bradley and his classmates, the only thought on Bradley's mind is planning a date with Beth.

Escorted by their assigned agents, Conley and Beth arrive just before the ceremony starts. The auditorium's filled with commotion right up until 3:00 p.m. Then, without introduction, the high school orchestra starts the music prelude, people find their seats, and the room quiets. Wearing her master's degree regalia, Beth sits with the faculty. Conley's dressed in a light beige shirt, no tie, khaki slacks, and a blazer. He finds an empty seat in the middle and the agents take positions around the auditorium.

Following the procession of the senior class and the invocation, Mr. Maxwell introduces the chairman of the school board. The program continues with a brief speech by the chairman and then Linda Evans, the valedictorian, steps to the platform. Her speech centers on college life and activities. Next, Mr. Maxwell announces several special awards before the school board presents the diplomas.

"For the first award to a graduating senior," Mr. Maxwell declares, "I'm pleased to offer a special science award. This honor goes to a senior who excels in science and intends to continue a career in this field. It's no surprise the award goes to Bradley Truman. Brad's won both state and national awards for his science exhibit."

Maxwell motions Bradley to come to the podium. As Bradley comes forward, the FBI agents reposition themselves. Their attentive posture is noticed by many in the audience and chatter arises.

"Don't worry," Mr. Maxwell says, pointing to the agents. "These folks are harmless. We're used to having them around." Putting his hand out to Bradley, he says, "Congratulations on this award. You put our school on the map this year." The audience claps as he returns to his seat.

"Brad's award," Mr. Maxwell continues, "is a ten-thousand-dollar cash grant from an anonymous donor."

Looking directly at Bradley, he says, "I suggest you give a portion of this award to your science teacher, Ms. Elizabeth Williams."

Turning back to the audience, "Thank you, Ms. Williams, for your timeless support and encouragement." Another round of applause emerges.

About an hour later, and not soon enough for the overheated crowd, the orchestra plays the Springfield High Alma Mater. Happy seniors march off the stage, down the hall, and congregate in the cafeteria. They remove their graduation caps and gowns and chat with each other about their family celebration plans.

"Bradley," Linda asks, "Do you have a special graduation party planned?"

"No. I think we're just having a small celebration. Your speech, by the way, was super. I especially liked the way you caught everyone's attention right at the beginning. What a great story."

"So who'll be your significant other at your small celebration?"

Bradley laughs. "I haven't invited a date. I'm linked to an FBI escort."

"Have you ever dated anyone?"

"I promised Mom I wouldn't date in high school."

"So your dating reluctance is about a promise to your mother? What about now? High school's over and we could harmonize together." She shows him a glowing smile.

"Well," he responds, "I appreciate the thought. Although I'll be dating sometime, I don't plan on bringing along an FBI chaperone." He points to Agent Matt Rodriquez. Giving her a guarded hug he says, "You've been a good friend. Best of luck."

Bradley makes his way through the crowded hallway. He meets his dad in the auditorium who gives him a hug. Fifteen minutes later, Beth catches up with them and gives Bradley a big squeeze.

"I'm so proud. Do you know who sent that award?" she says.

"I'm guessing AMC," Conley responds. "It might be an allowance for your time in San Diego."

"Well, whoever it was, I can now repay my savings account," Bradley says. "So, what's the plan?"

"How does a Jim Dandy sound?" Beth giggles.

"Let me check with Matt," Bradley cautions with a teasing wink. "Hey, Matt. Do you like ice cream?"

Matt tries to keep a professional look and doesn't respond, but his increasing grin supplies the answer.

"I think that's a 'yes,'" Reynolds declares.

"Okay, the six of us are going for Jim Dandy's," Bradley proclaims.

Agent Rodriquez steps up and hands Bradley the keys to the black Range Rover. "Since it's your party, you can drive the Darth Vader van. We'll all ride together. How's that for security?"

As Bradley drives to his favorite ice cream restaurant in Springfield, someone dials a phone number which extends halfway around the world.

"Frank, Vince here." A distinguished voice states.

"How did you find me?" Frank responds angrily. "I thought you would leave me alone."

"Can't do that. You're still on the payroll and the best man for my needs."

"They need me right here," Frank states.

"I need you in California for the summer. I'll send you more details, and you may even thank me for this assignment. Do you still talk with Alessia?"

"We talk from time to time."

"Well, I believe she's sitting on a time bomb that may blow up in her face. I don't want her buried by some EU dirty tricks. She's too important to both of us. I need you in San Diego before July first." The VP hangs up.

Chapter 13

THE LAST SATURDAY NIGHT of June, Bradley parks on a side street near Beth's apartment with a polished LTD. His week's been busy looking for a summer job, filling out overdue applications to his first choice colleges, and planning every minute of his first date night.

Beth's waiting for his arrival, but she's a little uneasy. After all the experiences they've had together, she wonders if Brad can treat tonight like a romantic date. She wants to be special, sweet-talked, and close enough to run her fingers through his hair, not discussing science.

Hearing the doorbell, Beth opens the door to the tall handsome man in blue jeans, a light green shirt, and a fashionable dark-colored windbreaker. She notices he carries an open corsage box containing a white orchid. She's speechless.

"May I pin this on your dress?" he asks politely. She nods with a wide smile. He carefully pins the flower to her fancy summer dress, a flounce sleeve smock dress with colorful splashes of floral designs.

Standing back he eyes the flower's position and smiles. "I've never seen you look so beautiful." Her twinkling eyes and glowing smile show her appreciation, but before she can form any response he adds, "I hope you're ready for a fun evening." She takes his hand, already charmed.

As they walk across the street, Beth says, "I'm glad we're together. How did you lose our escorts?"

"Let's pretend they're not around, but I haven't lost them," he nods towards a light tan Buick across the street. Reynolds rolls down the passenger window and makes a few discreet wave motions.

Arriving at the LTD, he opens the door for her. Once she's seated, he moves to the driver's seat and proceeds north to the outskirts of Springfield.

As they travel, they catch up on their activities since graduation. The conversation's light, cheerful, and Beth feels delighted.

Soon, Bradley pulls up to an old steam locomotive at a restored train depot. Although not dark, the engineer cab, as well as the seven train cars, are lit in an appealing pattern. The depot is surrounded with gas lanterns converted into electric lights. It's not busy yet, but a few customers are arriving and the staff's dressed in the attire of the late 1800s.

"Are we eating here?" Beth inquires.

"Yes. It's the Depot Station Restaurant. Do you like it?" Bradley asks as he maneuvers into a parking place.

"I've never heard of this place. It's like stepping into another world. Can we eat in a train car?"

"I think I can work that out." Bradley grins as he steps out of the car. He then moves towards Beth's door, but she's already out of the vehicle. They walk to the reservation desk, and Bradley asks for the Truman table. A train conductor cordially escorts them into an antique railroad car with about a half dozen tables for four.

Beth admires the polished oak wainscoting, purple window shades with gilded fringe, and dark green dome ceiling trimmed with gold. She moves aside her place setting, reaches across the linen topped table, and takes both of Bradley's hands.

"This place is amazing. I'm so glad you brought me here," she says with gratefulness. Bradley smiles from ear-to-ear.

For the next few minutes, they study the menu. Bradley orders an iced tea and Beth orders a coffee. She would have preferred a sweet red wine, but since Bradley's too young to order from the bar, she settles for the java. Bradley winks at Beth as their waiter comes to take their order.

"Please order whatever you want. That's what I'm doing." Noticing a slight reluctance in her eyes, he adds, "We both worked for that award, let's enjoy it."

Beth orders the lobster platter, soaked in butter; rice pilaf with almonds; and triple bean salad. Bradley orders prime rib, medium well; loaded baked potato, with extra sour cream; and steamed vegetables. As the waiter departs, Beth questions Bradley about the award.

"Do you have any idea who donated that money?"

"Dad did research on the check. The Government Employees Credit Union branch in Houston, the GECU, issued the check. That's as much info as he could get. It was a bank check, so no personal information is available."

"The GECU? That sounds familiar. I think my college tuition came from GECU checks. I always assumed my mom or dad, probably Dad, paid for my education."

"Is there some way you can check? Maybe your dad's still looking out for you."

Beth sighs. "I haven't heard from my father since he left. I doubt he cares whether I'm dead or alive."

A waiter seats another couple at the other end of the train car. He comes to their table and offers refills on the drinks. Since the conversation stopped, Bradley changes the subject.

"So what have you learned about San Diego?" he asks.

Beth beams as she shares all the places she wants to visit, starting with her top five choices: San Diego Zoo, Hotel Del Coronado, a Whale Watching Cruise, Cabrillo National Monument, and the Gaslamp Quarter. Then she begins with other locations in Los Angeles and San Francisco as Bradley listens patiently.

When the meal's delivered, they both inhale the delicious aromas. The waiter brings a bottle of champagne and two glasses. He pours for both and sets the bottle into an ice bucket near the table. Bradley can't help but snicker as Beth looks astonished.

"Did you order this?" she whispers.

"Yes," he responds.

"How'd you do it?" she says, again in a low voice.

"It's nonalcoholic," he winks.

They both laugh out loud until tears form in their eyes. Bradley calms himself, picks up his knife, and cuts into his steak. Beth takes her lobster fork, and, for a time, the only conversation heard are sounds of satisfaction. When the waiter comes to pick up their empty platters, they both send their compliments to the chef.

"Have you decided on dessert?" the waiter questions.

They look at each other and raise their eyebrows. Beth looks at the waiter. "There's no room for dessert," she says. The waiter clears their table and leaves to prepare the check.

"Brad, it's only 6:15. I hope you're not taking me home." Beth flutters her eyelids and provides a playful smile.

"I have plans," he says with a mysterious tone, "so we must leave soon."

Upon receiving the tab, he holds the waiter for a moment while he reviews the total. Once approved, he hands over his credit card.

"When did you get a credit card?" Beth utters with astonishment.

"As soon as I graduated. I'm getting with the program."

"What program?"

"The grow-up-and-be-somebody program," he answers. Calculating a liberal tip, he completes the receipt information and they exit the train car. Beth spends a few minutes perusing the railway depot souvenirs and then they're off again in Dad's classic.

Thirty minutes later, Bradley pulls into the parking lot of Calvary Temple, a large Springfield independent church.

"Are we going to church?"

"We are." he smiles.

The sanctuary is filling up, so Bradley hurries Beth inside to find a good seat. Looking across the seating, they spot ushers assisting the crowd. One usher points at two empty seats in the second row. They wave, rush forward, and sit in the plush church chairs.

A sound check is in process, and Bradley watches the sound crew in t-shirts and jeans. Beth's more interested in the instruments on the platform: a set of drums, electric piano, two guitars and lots of microphones. When the setup's completed, the sound crew disappears and an older gentleman steps out on the platform.

"Thank you for coming tonight. We apologize for the late start, but the airport diverted their flight to Cincinnati. With the extra travel, we're running a little behind schedule. But enough of that, let me introduce, from Santa Rosa, California, the Carlson Family Worship Team."

The piano player's first on the stage and begins playing a fast-paced popular Christian song. With the piano carrying the melody, the drummer, bass, and rhythm guitars fill in the music. As the audience claps along, five singers come to the microphones, and the church starts rockin' from one contemporary Christian song to another.

After an hour of non-stop hand-clapping music, the older gentleman again steps out on the platform. "The Carlson Family are traveling to Nashville to record their fifth CD. They agreed to stop over in Springfield for a freewill offering. Please be generous as we pass the buckets. I encourage you to buy any, or all, of their current CDs as well. We can show our love and appreciation by giving to their ministry. The second half of their program will begin in fifteen minutes."

"Brad, tonight's so special," Beth gushes. "This group's great. I can't remember when I've clapped and sang so much."

"I'm enjoying the music too. Do you want to stay for the second half? The worship time will grow in intensity."

"I'm game if you are. I need all the Spirit I can get. It's been a long time," Beth says and settles in for the next performance.

After the break, one of the Carlson Family members steps up to a microphone. "We appreciate the way you took part in the singing and clapping tonight. It's wonderful singing songs about Jesus and the Word. In this next set we pray the Holy Spirit will free you from your inner struggles. We want God's Spirit to touch your hearts, souls, and minds. Instead of using your energy to sing along, channel your energy into sensing the closeness of his Spirit. Worship takes us to repentance and the cleansing of our hearts. Let's bow our heads and ask God to speak to us."

As he prays, the team comes back on stage. The rhythm's slower, the lyrics more intimate, and there's an immediate sense of God's Spirit throughout the sanctuary. Bradley remembers services like these when he attended his mother's church in Columbus. He stands at his seat, raises his hands, and acknowledges the presence of God. With his eyes closed and his mind focused, he's not alert to Beth's emotional struggles.

Beth sits with her head down, eyes closed, and tears dripping on her dress. There's a perplexing look on her face, both humble and stubborn at the same time. With her mouth she begs for God's mercy, but with her soul she resists the conviction of the Spirit. She's only heard two worship songs, and she's ready to leave.

She tugs on Bradley's arm to get his attention, but before he can respond, the worship music quiets. "We'll continue to honor God with worship," the Carlson Family leader proclaims, "but if you're ready for prayer, you can come forward now."

"If I go forward for prayer, will you come with me?" Bradley asks.

"Can we leave after prayer?" Beth pleads.

"Of course," he replies.

They both move out of their seats and join with others approaching the prayer teams. When their turn for prayer comes, Bradley takes Beth's hand. They meet with a balding man and his petite wife.

"How can we pray with you tonight?" the husband asks.

"We have to make some difficult decisions," Bradley says. "We have a science project that has put our lives in danger. Our next step is flying to California for further testing of our invention. We need God's wisdom and protection."

"Are you two married?" the wife questions.

"No," Bradley replies, deciding that no explanation is necessary.

Turning to Beth, she speaks with concern. "You're not willing to forgive someone. Is it him?" She motions towards Bradley.

"I'm willing to forgive everyone." Beth responds defensively. "I don't have any issues with Bradley."

The husband and wife give each other a knowing look. They suspect Beth's in denial, but choose not to press the issue.

"Let's join hands," the man states. "We'll pray now."

After a short prayer, Bradley and Beth exit the church. Beth's annoyed and quiet, but Bradley continues with his plans for the evening. He exits the church's rear parking lot and drives outside of the city. In about ten miles he slows down and turns onto a dirt road. He makes his way up the twisting road and parks where they overlook the city lights under the stars. In the rear-view mirror he watches the lights of the FBI car park about one hundred yards away.

Any other time he'd prefer to be alone with his date. The escort car, however, gives him a sense of security. Their near death experience in Brussels is still fresh in his mind. He's thankful the FBI's serious about their safety.

Pushing back the bench seat, he releases the convertible top and motions to Beth to slide over beside him. He puts his arm around her as they gaze at the beautiful lights.

After fifteen minutes of silence, Beth speaks in a humble voice. "I'm sorry for the way I'm acting. I got all confused at the church service."

"It's okay. Sometimes God tugs at our hearts, and we don't know how to respond," Bradley answers. He waits a bit and says, "There's something I've wanted to say for a while, but waited until tonight. I love . . ."

"Stop!" Beth says in a frustrated tone. "Never say that to me." She slides back to the passenger side. "Every time someone says they love me, they leave me. Both my mother and father said they loved me and then disappeared. What kind of love is that? I'll never forgive them for what they said."

Bradley remains calm and waits for God's Spirit.

"That woman was right, wasn't she," Beth confesses as a tear slides down her cheek.

"Yeah, she was."

"What should I do?" Beth moves closer to Bradley again.

"I'm sure you know. Can you do it?" Bradley takes her hand. "God loves you and he doesn't leave. He's ready when you are." Bradley leans over and touches his head to hers.

Beth prays, "Lord Jesus, I should forgive my parents for leaving me. Give me your strength."

As she finishes those few words a dam of emotions burst. She sobs and hides her face in Bradley's chest. She tries to slow her emotions.

"Let it all out," Bradley encourages. "Wash away the heartache."

She pulls her face away, bends her head down, and sobs. Bradley holds her, realizing she's not aware that God's healing her broken heart.

Bringing herself under control she prays again, "I forgive my mom for leaving. I miss her, but I don't blame her. By your mercy, I want to see her again someday. I forgive my dad for abandoning me at college. I love him. I want to see him again, too. Forgive me for not trusting you." She takes his hand. "Thank you for the wonderful friend you've given me. He has expressed his love for me in so many ways."

Beth grabs tissues out of her purse and wipes her eyes. Bradley looks at her face and her relaxed expression puts him at ease. At that moment, his love for God and this woman permeates his emotions. Beth reaches out and strokes his hair.

"Bradley, each day you were in my class I wanted to put my hands through your curly hair," she smiles.

"Since the day we met, I've wanted to stare into your beautiful eyes, hold you tight, and kiss your lips."

She turns his face to hers. "Bradley, my sweet and faithful friend, please ignore what I said. I want to hear you say that you love me . . . over and over again." Gently, she puts her arm around Bradley's neck and places her lips on his. Bradley responds with a loving squeeze. As they embrace in their first kiss, their pent-up passion breaks free. They begin kissing each other's face and neck.

After a few moments, Bradley tenderly holds Beth's head against his shoulder. He sighs, "I love you." Then whispering in her ear, "I know you once wanted to watch the stars all night, but it's time to go. I always want you to feel safe with me, protected."

Beth calms herself and smiles. "Thank you. Thank you for loving me so much. I appreciate it. I really do."

Bradley starts the engine and raises the top as Beth leans on his shoulder. Once in gear they meander their way back to the main road. As they get partway down the landfill road, Bradley sees the other car lights following them.

Driving up to Beth's apartment, he parks and walks her to the door. With no hesitation, they put their arms around each other and kiss without shyness. After Beth closes the door, Bradley walks back to the LTD. The FBI Buick drives alongside and Reynolds rolls down her window.

"Where have you been?" Agent Reynolds shouts with a perturbed attitude.

"What do you mean?" Bradley responds. "You've been following me all evening."

Agent Rodriguez steps out from the driver's seat and stands at his car door. "Did you ditch us on purpose? We lost you when you left the church."

A terrified look appears on Bradley's face. "Then who followed us up the landfill? I thought it was you."

"It wasn't us!" Rodriguez says angrily. "You need to be more cautious. Your carelessness might have become life-threatening. We'll talk more in the morning. I'm leaving Agent Reynolds at Beth's and coming home with you."

The next morning Beth and Agent Reynolds meet at the Truman's home. Beth and Bradley are expecting a stern FBI reprimand, but Rodriguez and Reynolds surprise them with assorted bagels and coffees.

"We were a little too hard on you last night," Matt says to Bradley. "But, you must remember to double check your security. You'll be leaving for San Diego soon and stepping into unfamiliar geography. I don't want to scare you, but you can't trust AMC to do the whole job."

"I'm sorry," Bradley confesses. "I didn't intend to put us in danger. So it sounds like you won't be coming to AMC."

Agent Reynolds puts her hand on Beth's arm. "I love you both, but Matt and I will receive new assignments as soon as your flight leaves. The FBI will not assume any further responsibility."

"Well, what's this about a landfill?" Beth says with curiosity.

Bradley responds with a worried tone, "Beth, you once told me you wanted to go someplace where we could view the city lights as well as the stars. There're no mountains around here, so I drove to the top of a retired landfill."

"So you took me to a garbage dump on our first date!" she exclaims loudly with an unhappy look. There's a few seconds of anxiety in the room and then she giggles and says, "You're so sweet. You really know how to charm a girl." Everyone laughs.

For the next few minutes, the five share their appreciation for each other. When Reynolds goes to the refrigerator to refill the half and half container, Bradley follows her.

"Agent Reynolds," he asks, "have you done any research on Beth's parents? Where do they live? What are they doing?"

"Brad, Beth's mother lives in Oregon. She's remarried and works as a high school science teacher." She fills the milk container and closes the refrigerator door. "Beth's father is a US government chemist. He works on top secret classified projects all over the world. Last we checked he was in Australia working on artificial photosynthesis. That's all I can tell you."

"That's all you know, or that's all you can tell me?"

"That's all I can tell you," she says firmly.

"Is there any chance we could bump into him somewhere?" Bradley asks.

Reynolds stops and looks Bradley straight in the eyes. "Yes, and it may be sooner than you think." She grabs the container and walks back into the dining room.

Chapter 14

AS THE PLANE TAXIES into terminal one at the San Diego International Airport, Beth awakes from her snooze. It's Friday evening, and Bradley and Beth are ready for the next phase of developing their desalination invention. Earlier in the week, they each shipped a carton of clothing and personal items to the AMC lab, unsure of how long they'd be staying there. For the flight, they packed their luggage with a few weekend clothes and filled their carry-on bags with toiletries, Beth's laptop, and research papers.

"What time is it?" she asks wiping her eyes.

"It's 6:00 p.m. California time."

"That's when we left Cincinnati."

"Think of it," he laughs. "You slept three hours and the time never changed."

Traipsing through the airport, they follow the baggage claim arrows. Bradley stops at Starbuck's and picks up two venti iced mochas. Waiting at the baggage carousel, Bradley spots someone in a chauffeur hat walking around with a sign: Bradley Truman/Elizabeth Williams. Waving at the chauffeur, he captures the man's attention.

"I'll be your driver tonight," the man states. "As soon as we get your luggage, I'll take you to your hotel. Mr. Timmons wants you to enjoy the weekend. He says there's a lot of work in the next two weeks. He's arranged for you to enjoy San Diego first."

"Do you work at AMC?" Bradley asks. He's alert and cautious.

"No. I work for their security service. Here are my credentials, your security password, and an emergency phone number. You can call 911 to verify."

As the baggage wheel turns, Bradley steps aside and calls 911. He gives the operator his name and password. After a few minutes another dispatch agent joins the line.

"Mr. Truman, we're aware of your arrival. Please confirm your security agent's badge number is 458X." Bradley looks at the driver's credentials and confirms.

The dispatch officer responds, "You should be in good hands." Bradley disconnects, but keeps his mobile in hand.

"What's this emergency phone number?" he asks the driver.

"It's Paul Timmons' personal cell number. You can call him for any reason."

Bradley punches the number into his mobile and Paul answers.

"Hey, Paul," Bradley says. "Do you have my password?"

Paul chuckles and says "saltwater." He adds, "Thanks for being careful."

"Just checking," Bradley replies. "See you soon."

"Bradley, you and Beth have a great weekend. We did a little extra, so I hope you enjoy San Diego. Next week the accommodations won't be quite as pleasant. Glad you've arrived safely."

"Satisfied?" the driver quips.

"Absolutely, and no offense." Bradley insists.

The driver nods, picks up their luggage, and leads them to his car.

About thirty minutes later they pull up in front of the Hotel Del Coronado. Beth's excitement is obvious.

"Are we staying here?" she asks with delight.

The driver turns in his seat. "You'll be staying here three nights. I'll pick you up at 8:00 a.m. on Monday and drive you to the AMC Tijuana River Lab. Bring all your things, you'll be staying at the lab the rest of the time. Let me help with your luggage."

While making their way towards the hotel, Bradley and Beth can't help but marvel at the blue ocean view, the hotel's amazing turret, and the porches wrapping around the building. Catching up to their luggage at the reservation desk, Bradley gives their names.

"Yes, we have your reservation. You'll be staying at the beach village, number four. I'll have a bellhop take you and your luggage there right away."

Beth stiffens with a concerned look, but Bradley doesn't notice. The experienced reservation employee catches Beth's response and understands right away. He gets her attention.

"There are two separate bedrooms in suite number four," he affirms with a smile. Beth smiles in return showing her gratefulness.

The bellhop arrives, picks up the luggage, and leads them through the hotel to their beach village suite. As he opens the double entrance doors, they look right through the living quarters at the endless blue Pacific. As directed by Beth, the bellhop takes their luggage into two separate bedrooms. Upon his return, he puts out a hand. Bradley responds with a twenty-dollar bill.

After the bellhop disappears, Beth asks, "How did you know to tip him?" She playfully pushes him onto the couch.

"Because I know things," he responds pulling her on top of him. "I've got skills. And I've stayed with my dad at a few big name swank hotels."

Beth sits on his lap and teases. "I've got a lot of unpacking to do, but first I want a little lip service."

As they embrace, a knock comes at the door. Bradley looks through the peephole and sees the same bellhop.

Opening the door, the hotel employee says, "Ms. Williams, please." Beth hurries to the door.

"These are the keys to your rental car. It's a light green Chevy Cruise parked in our front lot. You also have prepaid tickets to the San Diego Zoo. They're good for Saturday or Sunday. All are compliments of Aquatic Mining Company. Just sign here for the car. When you check out, please turn in the car keys at the desk."

Beth giggles, closes the door, and throws her arms around Bradley. "Call your dad. Tell him we're not coming home." Bradley gives her a squeeze and a tender kiss.

"Good night my queen. I will unpack and get some sleep. I want to be on top of my game for you all weekend."

On Monday morning, security driver 458X transports them south for a twenty-five-minute ride. At the junction into the Tijuana River National Estuarine Research Reserve, they leave the suburbs behind and journey into protected wetlands. Continuing towards the Pacific they spot a large two-story white cinderblock building ahead.

"That's the AMC lab and your new home," the driver states.

As they arrive Matt and John Timmons are waiting in the parking lot to greet them. Upon shaking hands, Matt's all business.

"Now we'll get right to work, so let me give you direction. Bradley, you'll be living in apartment number fourteen. It's on the lower floor and your UPS boxes are already there. Robert Seibold, our CAD expert, is in number seventeen and he's waiting to start the drawings for the fabrication.

"Beth, you'll be living and working in apartment number twenty-three on the second floor. You'll be near John's office. Drop your luggage there and John will give you the regs, the California desalination regulations, to review. We need your evaluation on the regs right away."

Matt turns and hurries towards the building. As he swipes his security card at the main entrance, he yells, "I've got a conference call, so I've got to run."

"Matt's always in a hurry," John says apologetically. "Bradley, you and Bob should get along just fine. He's easy to work with. Beth and I will consider how well your system meets the regs." John slides his sensor card at the entrance door. "By the way, Beth, you already have a meeting on your schedule in about a half hour." Bradley and Beth gather their things and hurry to find their assigned rooms.

Bradley paces around his small apartment. There's an open area with a small kitchen, dinette, and couch. Two UPS boxes containing their shipped clothes are sitting on the floor. Opening a narrow pocket door, he finds a small bedroom with a twin frame and mattress. A light hangs on the bedroom wall and a metal bar protrudes near it. He concludes the bar is his closet. Next to the kitchen's a worn plastic folding door which reveals a sink, shower, and toilet.

He's about to carry Beth's UPS carton upstairs, when he sees a man in his late twenties standing in his doorway.

"Hey, I'm Robert, but I prefer Bob. You must be Bradley," he says with a cheerful grin. "Let's get started."

"Okay," Bradley responds. He tosses Beth's box on the couch, shakes Bob's hand, and follows him down the hall. "What's our goal today?" Bradley asks.

"First, you help me create several freehand sketches of your invention. We'll use the sketches to program our in-house CAD system this afternoon. Matt's eager to give our fabrication team the CAD design so they can build it overnight."

"You don't waste any time," Bradley laughs. As he follows Bob through the lower level, he passes other apartment doors. In a few more steps Bob opens a door marked 'FAB'. It's a large open space, but in one corner are four cubicles with computers, a plotter, and other high-tech equipment.

Beth takes time to unpack. Her apartment's much like Bradley's, but her bedroom has a little extra square footage with a dresser. Beth also has a large kitchen table wired for an office phone and laptop. She's about to put her things in her dresser when John knocks on the door.

"Your meeting's in five minutes in a conference room. I'll walk you over," John says.

As they begin down the hall Beth asks, "What's this meeting about?"

"One member of our staff insisted on meeting you right away. He's a chemist we hired from Australia. He says he thinks you're related. His last name is Williams."

Beth's stomach turns. She's not sure if it's fear or excitement, but she's suddenly nauseous. She opens her purse searching for an antacid.

"So what's his full name?" she asks nervously.

"Franklin, Dr. Franklin Williams."

Beth's heart begins to pound. She gets lightheaded and stumbles into John as they turn a corner. John puts his arm around her and looks her in the face.

"Are you okay?" he asks concerned. "Your face has turned white."

Her mind's overwhelmed. She's thinking, how? why? No one gave her notice, no preparation, and no hint of his connection to AMC. She breathes deeply to regain her composure and digs in her purse for an antacid.

"I'm okay," she replies bringing herself under control. "Can you ask Bradley to come upstairs?" She locates the pill and pops it into her mouth.

As John grabs a phone in a side room, Beth leans against the wall holding back her tears. She knows this is an answer to prayer, but this isn't what she expected.

Soon, Bradley's hurrying down the hall. She waits and hugs him. "It's time to meet my father and I can't do it without you."

As they walk into conference room 2E, her dad stands. He and Beth lock eyes. For a lifetime of moments, they stare at each other, feet glued to the floor, and without words. Tears stream down Beth's cheeks as she steps towards him. He responds with arms wide open and they hug with a long and tight embrace. As they hold each other Bradley waits patiently. Franklin looks his way.

"You must be Bradley," he says releasing Beth.

"Yes. It's a pleasure to meet you, sir," Bradley says and shakes his hand. He turns to Beth and asks, "Are you going to be okay?" She smiles and nods wiping her tears with her fingers.

"I'm fine." She takes his hand for a moment. "I know you have things to do."

"I look forward to getting to know you," Bradley says to Franklin. He exits 2E and heads down the hall.

Beth pulls back a chair from the conference table and sits. Franklin sits opposite her and they take each other's hands across the table top. For a long time, neither speaks, but a message of longing pours through their fingers. Franklin breaks the silence.

"After all these years, you're the likeness of my little girl. I've prayed for you every day, and I've prayed we would see each other again." He pauses and says, "I don't expect you to forgive me . . ."

"Dad, don't go there! I love you. I've missed you, and I need you. I've forgiven you and Mom for whatever happened. It would be nice, though, if someday you could be honest with me about your disappearance."

He looks away for a moment trying to control his emotions. "I can tell you this. While you were in elementary school, I took a job from an old friend I met in my Peace Corps days. He asked me to engage in some very important, but dangerous, work. I foolishly agreed. Your mom was heartbroken by my decision. She was fearful I put our family in jeopardy. After a few years, I was in too deep. She left begging to take you with her, but I wouldn't let her. She left with nothing: no planning, no income, no family members to help her, I couldn't let her take you.

"By the time you turned sixteen, I was over my head in threatening situations. Since you were gifted enough to go to college, I decided it was time to disappear. I was always afraid someone might hurt you to retaliate against me. I'm so grateful to God he's protected you over the years."

"Dad, thank you. I always knew you loved me. There's so much I want to say, but I don't know where to start. Bradley's the greatest thing that's ever happened to me. Please get to know him."

"I'm sure we'll be working together. I have a car, so we're not stranded. We'll get off campus a few times each week, just the three of us."

John sticks his head in the door. "Are you guys doing okay?"

Beth walks around the table and puts her arms around her dad's neck. "John, this wonderful man's my long-lost father. We haven't seen each other in almost ten years. It shocked me today when you said his name."

"Well, I didn't have a clue. I'm sorry I couldn't have better prepared you. Would you like a few days to catch up?"

"That's very kind," Beth replies, "but I'm not ready. I can't even think straight."

"It's okay, John. Thanks, maybe another time." Franklin stands as he answers.

Beth puts her arms around her dad. "Can we talk more at dinner?" she asks. Franklin nods and kisses her forehead. Then turning to John she says, "I'm ready to read those Cal regs. I need something boring to concentrate on so my emotions will settle down."

Chapter 15

THAT AFTERNOON, BETH TRIES to concentrate on California desalination regulations. She's still quite emotional because of the surprise family re-union. After making several notes about the water intake regs, she walks down to the first level and looks for Bradley.

She steps into a large open room referred to as the fabrication room or FAB. Bradley's office sits among the four cubicles. As she looks for Bradley's desk, she hears someone in a cubicle dialing a fax machine. She rounds the corner and meets a stranger.

"Hello, I'm Beth Williams," she says boldly. She extends her hand as she speaks.

She catches the person at the fax machine by surprise. "Oh, hey, I'm a chemist," he responds with a European accent.

"What's your name?" she insists.

"I'm . . . Tobias Gruber," he replies. "I'm just leaving."

"I'm looking for Bradley Truman. Do you know him?"

Tobias looks uneasy. As the fax machine successfully feeds the first of several pages, he says, "I must go now." He hurries towards the stairs.

Curious about his suspicious behavior, Beth peeks at the document being faxed. In an instant, she realizes its hand drawings of their desalination system. Faxing this material doesn't seem right, so she carefully fingers through the remaining pages and removes the sheet illustrating the wire loops. Grabbing her phone, she takes a photo of the digital fax number. When the fax pages are completed, she reinserts the missing page and waits for Bradley.

"We've been outside," Bradley admits, strolling into the FAB cubicles.

"I like to have a smoke once in a while," Bob grins, "but he wouldn't join me."

"Why would you be faxing these drawings out of the lab?" Beth asks, pulling the document out of the fax tray.

Bob grabs the document and sits down, mouth open, and red faced. "Oh, my god! Do you think someone faxed this document?"

"I just took it from the fax tray," Beth exclaims.

Bob looks at Bradley, "Where did we leave this when we went outdoors?"

"I thought we left it on top of your desk." Bradley retorts. At that moment, the fax machine pumps out a verification page marked "successful."

"I must report this to John. This looks like a security leak. But first, we've got to finish the CAD specifications. I'll talk with John in the morning." Turning to Beth, Bob says, "Thanks, but I'll likely get fired for this screw-up." Bradley follows Bob to the CAD computer and gives Beth a thumbs up and a light caress on the arm.

Beth goes back to her apartment and sits for a minute. What a day, and it's only half over, she thinks. She's upset and digs to find another pill. She sets the regs aside and dials Conley.

"Conley, I think I've discovered a spy here at AMC. What should I do?" she whispers into her mobile.

"Are you sure?" he says with skepticism.

"I caught someone faxing our desalination drawings to a foreign phone number."

"Okay, okay," he pauses for a minute. "Why don't you call Mark Conti? He'll know what to do. I'll text you his office number."

Beth becomes frightened. She closes her apartment door, moves into the bedroom, and locks the door. She explores the room for a microphone or camera, but sees nothing. Popping a chewable into her mouth, she dials Mark's office number.

"Hello. United States Patent and Trademark Office," Missy Kearn announces.

"It's Beth Williams. I urgently need to speak to Mark."

Seconds go by and Mark answers. "Hey, Beth," he says cheerfully. "Are you in San Diego?"

"Yes. I think I found a spy," she whispers.

Mark's quiet for a moment.

"Mark, are you still there?" Beth frantically pleads.

"Sorry, yes. I've invited Missy to join me. You're now on speaker and I'm recording. Now tell us exactly what's happened."

Beth explains her encounter with Tobias.

"Can you text the fax number?" Mark requests. Beth keys the number in her phone.

"Beth, give me his name in the text as well," Missy adds. "I'll contact AMC's security company and get his background report. They should have taken his prints in case he's using a false name."

"Beth," Mark explains. "You can contact Missy directly at any time. She's a member of the FBI unit investigating the desalination plant disasters and related murders. I'm sorry to say we learned Joshua was leaking confidential information. Maybe that's why he was killed, to sever that loose end. It's all very distressing."

"Where are you calling from?" Missy questions.

"From the bedroom of my apartment at AMC."

"Please don't call me from inside the building again. We can't trust AMC's security." Missy stops for a minute and then says, "I just got your text. I'll do a little research and get back to you."

While Beth's on the phone, Bradley and Bob are completing the initial CAD input from the sketches. After finishing the information, they double check every input and consider additional specifications. Bob's usually a perfectionist regarding his designs, but Bradley senses his greater concern is losing his job.

"If I'm still working here tomorrow, we'll meet with the chemists to get help with the nickel plating specs. We should ask about your twelve-volt battery system for the loops. Not sure if twelve or one hundred twenty volts makes any difference."

Beth and Franklin meet for dinner in AMC's small cafeteria. There is only a half dozen interesting choices in the vending machines, but Beth's not hungry. She's excited to tell the story of Brussels. Although careful not to disclose all the details, she goes through their trip from start to finish. Franklin's amazed at the danger they experienced. When they're alone, she also reveals her meeting with Tobias and unloads all her fears. Conversing back and forth, their father/daughter bond rekindles.

Bradley and Bob finish the fourth double check of the desalination specifications. Bradley skips dinner and heads back to his room, but he stops by the cafeteria to say goodnight to Beth and her dad. He's ready to take a shower, get settled, and crash for the night.

The first night in a new bedroom and on a well-used twin mattress, it's difficult for Beth to sleep. As she rolls around on her bed, she hears a constant tap, tap, tap, at her door. At last, she gets up and cracks open her apartment door. Bradley and her father are standing in the hall. Before she can speak, Bradley puts his finger across his lips. After a few unorthodox hand signals, they exit the building, walk into the parking lot, and follow Franklin to his maroon Buick sedan.

"We still need to be quiet," Franklin whispers as he demonstrates a quiet way to open his car doors. Once inside the car, he explains with a low voice.

"I still have ongoing responsibilities in Australia, so I come out here in the middle of the night and phone my Australian colleagues. When the security company checks the entrance door logs tomorrow, they'll see I've exited and reentered the building. However, they'll assume I'm alone and calling my Australian team."

"So what are we doing here?" Beth asks.

"Tonight I'm calling an Australian number, but I'll be speaking to someone in DC." He dials the phone and puts it on speaker.

"Is this the doctor?" a voice answers.

"Yes, me and my team."

"Someone on your team contacted us. The person in question is who he says, and he's on our 'wunderbare' radar. He's also connected to offices in Bern. Don't. I repeat. Do not interfere with his assignment. Observe his activities, but be careful. Do you understand?"

"Who's our best contact? What are the next steps?" Franklin asks.

"The office you called today is your POC. I'm planning to arrange a new venue, but don't shortcut your responsibilities. The Italian will contact you. Let me know immediately if you cannot meet her request. And doctor, it's time to be up front with your team. I enjoyed meeting them. Thanks for your assistance." The phone disconnects.

Franklin checks his watch. "We've only got about five minutes, so we may have to finish this conversation another time. No questions now, I'm transmitting, not discussing." Bradley smiles at his comment.

"About fifteen years ago, I joined the CIA. My reputation in chemistry got me into places where Americans weren't always welcome. My handler at the time was Kennedy Kleaver, now the vice president. It was him on the phone. His meaning of POC is a point of contact. The Italian referred to is Alessia Amato. The 'wunderbare' term refers to Wunderbare

Flüssigkeitszufuhr. Wunderbare was a fly-by-night water bottling plant in Switzerland about twenty-five years ago. They were a violent lot and created trouble throughout Europe. We believe those same people are causing trouble now.

"We need to go. A typical phone call only lasts about ten minutes. Let's try to open and close the car doors at the same time. We'll talk more when we can."

In the morning it felt like she'd never slept. After two cups of coffee, Beth readies herself for work and opens her apartment/office door. Franklin pops by and gives her a kiss on the cheek. John knocks on her open door about 9:00.

"So, my biggest concern's the seawater intake regs. How does your system line up with the requirements?" he says pulling out a chair.

"In our model, the water flows under one mile an hour. Slow water allows more time for the molecule separation. I believe we can avoid many of the regs that protect the fish and shoreline. California wrote these regs for higher speeds."

John takes a breath. "Good news, but let's be 100 percent sure. So how is your reunion with Franklin going?"

For the next several minutes, Beth and John share their favorite childhood memories of their parents. The conversation continues until Bob knocks on the door.

"John, I need to discuss something with you. Do you have time this morning?"

"Of course. Do we need a private office?"

"It's about a security leak," Bob sighs. "I left the drawings of the desalination system on my desk and I think someone faxed them out of the country. I take responsibility. Beth knows about it."

"Have you checked the outgoing fax numbers?"

"Yes, all are standard numbers except one to Switzerland."

John turns to Beth. "And you know about this?"

"John, this is a misunderstanding," Beth responds. "Bob, I'm sorry I caused you such distress. I got overexcited at the fax machine. Your sketch document was not part of my fax to Bern. I picked up your drawings by mistake. I thought Bradley straightened it out with you."

John looks at Beth and then Bob. "So we're okay, then?" he asks.

"We're just fine," Beth replies. "Thanks for your honesty, Bob."

John and Bob exit Beth's room and she turns her attention to the brine regs. Under her breath she mutters a prayer, "Forgive me, Lord, I just told a lie."

Back at his office, Bob pushes his chair next to Bradley. "I think Beth just saved my job," he sighs.

Bradley puts his hand on Bob's shoulder and whispers, "The Lord works in mysterious ways."

After a brief lunch, Bradley and Bob gather the AMC chemists in a large conference room. Bradley explains how the nickel plating functions to the group. Franklin volunteers to work with Bradley on developing the right specs for the nickel. Tobias asks to join based on his experience in molecular partitioning. Franklin's cautious of Tobias, but agrees. He recommends the three move to a different room while those remaining consider other parts of the system.

After they leave, Bob explains the wire loop and its function. Two chemists with industrial experience volunteer. The rest of the scientists return to their tasks of improving the brine for resale. By the end of the afternoon, both special teams are deep in debate over the millimeters for plating and the volts for wire loops.

After the evening meal, the six meet in the fabrication room. Beth joins them. Reviewing the CAD drawings, she's overwhelmed by the size of the system. The core stretches through most of the large lab, and the fabrication crew's already produced several pieces of the finished product.

Seeing Tobias in the room she deliberately strikes up a conversation with him. She wants to distract him from eavesdropping on the wire loop discussions. Although it's a late night for everyone, Bradley's invention is coming alive. He is too thrilled to sleep.

The first week's almost over when Beth gets an international call Friday morning.

"Hello, hello," a woman's voice says.

"This is Beth Williams; may I help you?"

"Oh, Beth. Alessia Amato here. I hope you don't mind me calling. I wanted to speak to you privately before any official call.

"When your dad and I worked in Africa, we became very close. When our Peace Corps assignment ended, he proposed to me, but the world was too big in those days. I couldn't leave my home in Milan and . . . I think, you understand.

"Anyway, I'll be calling you officially to come to Europe. I want you to build a desalination system for a town in Germany. If you think it's too awkward for us to work together, just say no. It's okay."

"Alessia," Beth asserts, "we're preparing to initialize our system here next week. We don't know yet how well it will perform."

"Okay, I understand, but this is about politics," Alessia replies. "People here don't care how well it works. I just didn't want our relationship to be uncomfortable."

"Thanks, but we'll get along fine," Beth responds. "Do you and Dad ever talk?"

"We have a friendly call every couple of months. He lets me know where he's stationed. I would like to talk to him more often. I miss him."

Suddenly, an alarm rings throughout the building. Beth sees employees rushing past her apartment door.

"Sorry, I've got to hang up!" Beth says going towards her door. Bradley runs in and takes her hand. They run out of the main entrance and meet up with the other two dozen scientists and professionals in the parking lot.

A SWAT team van arrives followed by a half dozen police cars and two FBI vehicles. A policeman orders everyone to the back edge of the parking area. Dogs emerge from the SWAT vehicle and, with K9 officers, rush into the AMC building. An FBI leader pulls Matt and John aside. Everyone's tense. No one seems to know anything.

After twenty minutes a policeman and dog emerge from the front entrance. He's carrying a medium size FedEx box and gives a thumbs up sign. A squabble takes place, and everyone turns to see the FBI putting Tobias in handcuffs. He's shoved into the back seat of a black Range Rover and driven away. As the police leave, Matt gives an "all clear" sign and everyone reenters the building.

At noon, a Domino pizza delivery truck drives into the parking lot. The driver, and his helper, carry a dozen large pizzas to the cafeteria. Bob summons Bradley, John collects Beth, and soon all the employees are in the cafeteria. Even Paul Timmons has arrived from his San Diego office.

"We've survived an unprecedented event this morning," Matt begins. "I'll explain what I can in the next minutes, but help yourself to the pizza while it's still hot."

As the employees crowd around the table of assorted pizzas, the facility manager's directed to unlock the soda vending machine. Each employee helps themselves to a few slices and grabs a free bottle of water or a carbonated drink. Beth and Bradley pick up their favorites and sit with Franklin.

When most everyone is seated Matt stands. "Last night search dogs, at the FedEx facility outside of National City, picked up the scent of explosives in an international package. When they scanned the package, it showed wiring, detonators, and other explosive material. They contacted the FBI, but the FBI requested the package go ahead as intended. The risk of an explosion was low since the scan proved the contents were not connected. The package arrived here this morning.

"Our chemist Tobias Gruber was the recipient of the shipment. He came to us highly recommended and vetted by our security team. The only suspicious activity we've monitored are two recent international calls to the same overseas office. One of those calls occurred when Tobias was seen using a FAB fax machine.

"The police and FBI followed the FedEx delivery truck here this morning. The lead agent told me there were enough explosives to destroy this entire facility.

"What we do here has great economic and political significance. We cannot assume this kind of aggression won't happen again." Matt pauses for a moment and adds, "So, if any of you wish to leave, I'll cancel your contract and pay you through the end of the month. Don't feel embarrassed. Do what's best for you and your family."

It becomes very quiet. Everyone looks around, but no one asks questions or leaves the cafeteria.

"All right. Now for the good news. As you know, we are building our own desalination system. We'll complete the fabrication this weekend and next Monday we'll give it a shakedown."

Everyone cheers and people return to the pizza table for extra helpings. Franklin pulls Bradley and Beth aside.

"Let's plan to drive to La Mesa for dinner. When Paul interviewed me, he took me to a real nice bistro. I'd love to take you. My treat tonight."

Chapter 16

SATURDAY MORNING BRADLEY GETS up early, grabs a cup of coffee, and hastens to the FAB area. He's amazed seeing his creation partially assembled. The IronWorks' frame was about eight feet long, but the AMC prototype will be over forty feet. Bradley reminds himself the forty feet doesn't include the water pipe from the ocean. Once a few pieces of old rain gutter and two rusty fifty-gallon drums, it is now in fabricated steel, Plexiglas, and sixteen-inch steel pipe. Realizing AMC's investment in this project, he fears the system may not produce the fresh water and brine promised. He turns on Bob's computer and studies the CAD drawings over and over scrutinizing for errors.

Bob arrives about thirty minutes later and spots Bradley pouring over the system design.

"You're scared to death," Bob laughs.

"And you're right!" Bradley exclaims as his face turns red.

"Well, let's try a software patent application. First, you input key facts about your invention. Then the program asks questions necessary for a patent filing. I realize your patent's started, but this may jog your memory in case we missed something."

Bob and Bradley sit together, pull up the program, and begin inputting basic system information. After twenty minutes the program starts asking patent related questions. The first couple questions are easy and Bradley types in the requested information. The third question isn't so easy: *What's the optimum water temperature?*

Bradley stops.

Bob glares at him asking, "Well, what's the temperature you used in your demonstrations?"

"I don't know," he shrugs.

"How high did you set the microwave unit?" Bob probes looking baffled.

"I just turned the microwave power on number seven and the water temperature worked."

"Duh," Bob grins pointing at the new prototype heating unit. "We need an actual temperature."

"I've got to look at my notes," Bradley says getting flustered.

"Let's go through more questions," Bob suggests. "Maybe there are other facts we need from your notes."

The next question from the patent program asks the measurements of the nickel plating, both length and depth.

Bradley again responds in anguish. "The depth question I can answer, but I never measured the length of the plating. I'll look like an idiot on Monday."

"All true inventors have idiot days," a man's voice says. Bradley looks around to see Beth and Franklin.

"Don't worry about the nickel plating," Franklin adds. "That's in my court. Want a late breakfast with Beth and me?"

"I guess I'd better stay here. I'm learning what I don't know. Have fun," Bradley says trying to smile, but feeling foolish.

Beth gives Bradley a peck on the cheek. "I'm sure you boys have everything under control."

Franklin drives to an out-of-the-way restaurant overlooking the Pacific. As they're seated next to a large picture window, the mid-morning ocean view is fabulous. The sun shines above the water which calmly laps against the shore. Both order black coffee and search through the breakfast menu. After the first sip of caffeine, Beth speaks.

"Dad, we haven't talked about Mom," Beth says cautiously. "I'd like to know about her. Do you ever talk with her? Have you seen her since the divorce? Does she ever ask about me?"

"Whoa, slow down," he responds. "I'm just having my first cup."

The waitress takes their order and tops off both mugs of caffeine.

"I guess the best way to describe your mother and my relationship," Franklin begins, "is she never felt loved by me. She believed she was my second choice bride. It began to affect our marriage about the time you started kindergarten. She always assumed I lost my true love and settled for her."

"Was Alessia your true love? Did she know about her?" Beth inquires matter of fact.

Franklin's stunned by her questions. At first, he's angry. He wonders how this information is known by Beth. Soon, however, his expression turns to a sad gaze, and he confesses.

"You're right, Alessia was my true love. When I accepted the fact that we couldn't be together, I came back to the states and enrolled in graduate school. I met your mom there. We dated for a long time. I know I cried on her shoulder a few times about Alessia. We were foolish to get married."

Franklin stares into Beth's eyes. "Who told you about Alessia?"

"She did," Beth admits. "We talked on the phone. She said you proposed to her years ago, and she misses you."

Franklin sits quietly. He pushes his food back and pulls his coffee forward. He stares at the Pacific as Beth studies his uneasy reaction.

"Dad, I'm sorry," she says and gently puts her hand on his arm. "I didn't mean to upset you."

He looks at her while nervously turning his coffee cup. "Quite the contrary. You said she misses me. My mind just wandered back. I haven't seen Alessia in over thirty years."

"Isn't she part of our assignment?"

"Yes, but I didn't think I'd see her again. Perhaps I should call . . ." His voice trails off as he speaks. He turns his head again towards the ocean. Beth lets him ponder and finishes her breakfast in silence.

It's just after lunch when they arrive back at the AMC Lab. As Franklin parks the car he gets a call. It lasts fifteen seconds. He turns to Beth.

"Vince wants to meet us at 6:30. You and Bradley need to come."

"Who's Vince?" Beth asks.

"Our contact's code name."

"I can get Bradley. Where are we meeting?"

"At the San Diego Zoo." Franklin smiles.

"You joking?"

"No. I'm not joking. We need to be ready by 5:00."

As they walk into the FAB, Bradley and Bob are examining the prototype construction. Bob's checking the intake and exit channels and many of the chemists are examining the brine reservoirs.

"Hey, I got the temp," Bradley exclaims to Beth.

"How?"

"Dad, took it in Kansas City. Someone asked him how hot the water was and he didn't know. He went around to other exhibits and borrowed a F&C temperature gauge. He noted the water was one hundred forty-six degrees Fahrenheit and about sixty-three degrees Celsius."

"Anything else we missed?" Beth asks.

"The length of the nickel plating and the wire gauge size for the loop. I think we'll work that out next week."

Franklin pats him on the shoulder and says, "That's good work. Let's go out tonight and celebrate?"

"You'll never guess where we're going," Beth asserts. "We're going to . . ."

Franklin clutches her arm. He gives her a serious look. "Let's surprise him."

"Oh yeah. Sorry," she sighs.

The San Diego Zoo closes to the public at 6:00 p.m. However, at 6:30 the vice president; his youngest daughter and her husband; and five grandchildren arrive for a special guided tour. As Franklin, Beth, and Bradley enter the park, secret service agents verify their IDs. They're searched and directed to an underground break room used by the Zoo's staff. Walking down several steps they reach a cinderblock room painted in rainbow colors. There are several tables with chairs, five soda machines, and eight different sandwich and snack dispensers.

As they grab chairs, an attractive FBI agent greets them.

She smiles at Beth and Bradley and says, "I know you, but you don't know me. I'm Missy Kearn from the Germantown USPTO." She shakes Bradley's hand. "I've worked with your dad."

Beth introduces her father and the three listen to Missy's version of the assault on their office. As they talk, a small group of international executives join them escorted by men in foreign military uniforms. Franklin instantly recognizes Alessia among the group, but she doesn't recognize him.

Without announcement, Vice President Kleaver enters. Everyone stands, but he motions them to sit. With no welcome pleasantries, he gets right to business.

"This is now Operation Saltwater," he declares. "It's not about climate change. It's about stopping the destruction of desalination technology and the murder of innocent people. Someone's targeting desalination innovation and personnel. We suspect it's a scheme to safeguard the European water bottling companies. Let me introduce our team's key players.

"The primary contact state side is FBI agent Missy Kearn. She needs to be in the loop on anything related to the US." He asks Missy to stand.

"Lina Martens' her counterpart on the other side of the pond. She's Alessia's admin and an Interpol intelligence officer. She's also a double agent, so be careful with your communications to her." He points to Lina, and she stands.

"The front line team in the US is Franklin Williams; his daughter, Beth; and Bradley Truman, the inventor of a new desalination technology being tested. Will you three stand please?"

"Franklin," Alessia exclaims, "I didn't recognize you."

The VP smiles and continues, "And that's Alessia Amato. Alessia's the leader of Eau Suprême, an organization protecting European water reserves. She's appointed by the European Union. Let's take a minute to shake hands and get acquainted."

Alessia makes her way to Franklin, kisses him lightly on both cheeks, and puts her arms around him. "Frank, it's been years," she whispers, "I've missed you."

Then nodding towards Beth, she says, "I heard your daughter speak in Brussels. She's a brilliant scientist like you. You must be very proud."

After hand-shaking introductions and light conversation, the VP asks for the group to sit again. He pulls a folder from his briefcase, clears his throat to gain attention, and begins his briefing.

"We've long suspected Finn Schweitzer, a minister in the Swiss government, is behind the damages and murders. He has a violent history, but I don't want to waste time sharing his background. Let me tell you what we know to be current.

"Witnesses in South Africa identified Karl Muller in Nhlabane when a desalination plant was put out of service and four people were murdered. Karl's been a favorite enforcer for Finn over the years and wanted by Interpol. Through the efforts of Hans Guttmann, the California Water Assurance plant was bankrupted. It was Finn who recommended Hans to them and we have a phone recording to prove it. Using fingerprints, we determined at least one assassin at the Groen Internationales Hotel, where Beth, Bradley, and others lodged, is part of Finn's old gang as well as those who attacked the USPTO in Germantown. A man named Alec, whose surname we don't know, had a hand in the most recent desalination plant explosion in Ras Al

Khair. He carelessly left his cell phone at the site, and we traced several calls to Finn's office. And just a few days ago the FBI successfully stopped a plan to blow up the Aquatic Mining Company here in San Diego.

"Our purpose is straightforward. We intend to discredit Finn in his own backyard. We'll bring Bradley's new technology to Europe and ask Finn to promote it. He'll either expose himself, or his supporters will turn against him. No matter, we need to act quickly and stop this man.

"Now, there are two wrinkles to consider. First, we've evidence that Diego Gonzalez recommended the chemist who tried to blow up the AMC. He's Alessia's supervisor and one of the five presidents of the EU. We're not sure about his allegiance. Second, based on a phone call from Alessia to Beth in Brussels before the assassinations, someone's advocating the idea Alessia knew about the Brussels hotel attack and warned Beth." The VP lets the information sink in.

"Okay," Vice President Kleaver says. "Frank, I need to know how fast you and your team can train people to set up the desalination system in Europe. Alessia, I need a suitable location and your signal when you've convinced Finn to promote it. We must be careful to get Diego's buy-in if it's needed."

Beth and Bradley stand at the same time. Beth looks at Bradley and lets him speak. "With all due respect, Mr. Vice President, we should be ready by the end of this week. There's no need to train anyone else." They stand waiting for the VP's response.

Franklin looks at Beth and exclaims, "This operation is for trained agents, not for two kids barely out of adolescence. I can't let you go on a dangerous mission."

"Frank," Alessia pleads. "In Brussels they were targets and at the AMC almost blown to bits. You can't protect them."

Bradley speaks again. "If our tests at AMC go as well as we expect, we're the best-trained personnel for the job."

"I agree," Kleaver responds. "I hoped you'd feel that way. We'll see where we stand at the end of the week. Thanks to you all for coming."

Immediately the vice president and his staff exit. Alessia tugs on Frank's arm.

"Come to Europe. I'll introduce you to my team. We can spend time together." She gives him another European kiss on both cheeks and slips something into his pocket. Alessia and her staff exit, leaving Franklin, Beth, and Bradley alone in the break room.

"This week's been crazy," Beth moans as they drive back to the AMC Lab. "I need to go to church tomorrow. Dad, have you attended any local churches?"

"Yes. The first week I was here, I went to church with Paul and his family. It's a charismatic church where things get loud and active, but I enjoyed the congregation's participation. The Pastor gave a great sermon. I wish I could remember what it was about. We can go to their 11:00 a.m. service."

"What do you think, Bradley?"

"I have no problem with loud and active. Remember my background's Pentecostal," he chuckles.

"I need to apologize for a lie I told to Bob and John," Beth responds. "I hope I can do that before Monday morning."

Tired, she lies down on the back seat as they drive to the lab. She looks up at her dad and Bradley sitting in the front seat talking. She whispers to God, "I'm so grateful for your love. I could never have guessed two such wonderful men would love me." She closes her eyes.

Chapter 17

Monday morning, the fabrication room fills with employees. Matt's on a conference call, so everyone waits for his arrival to start the official system testing. Beth corners Bob and John near Bob's desk.

"I need to confess I lied to you last week," she starts humbly. "I saw Tobias fax the system sketches, and I lied about my actions."

"Why lie?" Bob asks. John listens intently.

"The FBI did not want to raise suspicion about Tobias. They wanted to know if there was more to his assignment. I guess there was."

"Thanks for your openness," John says seriously. "We want to trust you and Bradley."

"How'd you know what the FBI wanted?" Bob questions.

"You know the saying," Beth grins. "If I tell you, I'll have to shoot you."

At that moment, someone comes running down the stairs. Everyone looks attentive thinking Matt's ready to kick things off, but it isn't Matt. Conley enters. He surprises Beth and Bradley, and a group hug ensues.

"Dad, it's so great to see you," Bradley says with a huge grin. "What's the occasion?"

"This is a big day," Conley replies, "for you and," he grabs Beth's hand, "us. I thought I would be late. The cab driver couldn't find the Tijuana Reserve entrance road."

Franklin shakes Conley's hand and says, "I've been looking forward to meeting you."

Matt comes hopping down the stairs and makes an attention-getting skip dance across the floor. Prancing past the employees, he can't keep a smile from spreading across his face. He's filled with entrepreneurial self-satisfaction.

"Okay," he says. "This is great. This is wonderful. Let's get this show going."

Looking around the room, he calls, "Bob, up front with me."

As Bob walks towards the front, Matt demands, "Tell me the fabrication's 100 percent finished and we can make brine."

Bob responds with an awkward look. "Well sir, no and yes. The fresh water retrieval system isn't complete. Once we identify a purchaser's requirements, we can complete that part of system. Right now fresh water flows into the wetlands. But, the rest of the system's ready to produce brine."

Matt responds, "That's want I wanted to hear. Take me through it."

Bob waves to Bradley to come assist, but Matt pulls on Bob's arm. "Bob, Bradley's leaving soon. This system's your responsibility now. I want you to take me through it, but keep it at a fifty-thousand-foot level."

Pointing at the large Plexiglas reservoir tank on the left, Bob begins, "The saltwater comes in at the top of this tank. When the tank is 90 percent full, it flows into the system channel."

Matt interrupts. "What's the water speed?"

"The pump has a maximum speed of five miles per hour. However, it's likely we'll run closer to one mile an hour, at least for the shakedown."

Looking at John, Matt asks, "Does that meet Cal's regs?"

"Yes, the regs do not take effect until water's moving much faster. We've dodged a lot of extra inspections and expense." Matt gives John a thumbs up and motions for Bob to continue.

"The reservoir catches silt that gets through the intake screens and will deliver clean saltwater to the system. Once flowing, the alteration of the water molecules happens here." Bob points to markings on the fabricated pipe which depicts the location of the interior nickel plating.

"The molecules move to the wire loops at this point," he says pointing to another pipe marking. "Then the brine falls into the right side tank."

Stepping to the brine tanks, Bob sums up the process. "These two tanks will hold different concentrations of the brine. These water taps extract the brine at different levels so we can sample different ppms."

"Did we get the nickel plating and wiring apparatus resolved?" Matt asks.

"Franklin developed a nickel plating ramp. It starts with 0.01mm and ends at 0.06mm. As a result, we lengthened the nickel plating from the original thirty centimeters length to sixty. The voltage debate ended with a rheostat which adjusts the voltage from twelve volts to one hundred fifty volts."

"Sounds like we need to test those two items this week," Matt states. "Now, what are we calling this contraption?"

Bob smiles ear-to-ear. "I've got a name. I recommend the Brine Reclamation and Desalination system. In short, the BRAD." Everyone looks at Matt, waiting for his response.

Matt looks around the room and makes a happy face. "I love it." Everyone claps and several pat Bradley on the back. "Give me a status report the end of today," Matt says and heads back upstairs.

Bob strolls over to the control panel and starts the pump. The BRAD's ready for testing. Two of the fabricators, the wire loop chemists, and a quality control inspector stand by to observe.

Conley tugs at Bradley's sweatshirt and says, "Get Beth and her dad. I've got exciting news." Bradley, Beth, and Franklin take Conley upstairs and find an empty conference room.

"Mark Conti called me yesterday to say they filed your patent. Apparently the CAD software you sent him included everything he needed," Conley beams. "I'm so proud of you both."

"Doesn't someone at the patent office have to approve the application?" Bradley asks.

"No," replies Conley. "Mark says they consider a patent complete at filing. However, the USPTO will eventually review the documentation. Mark says he deliberately left out the wire gauge specification for the loop. He'll know when the patent is being reviewed when the USPTO calls to request that information.

"The patent's not only good in the US but also in countries where we have reciprocal patent agreements like Canada, Mexico, United Kingdom, and all European Union countries. This is an amazing achievement."

"Dad, it's great to have you here. This means you can complete the license deal."

Overseas Brussels is cloudy and cold. A light rain's hitting Alessia's office window and she's frazzled over her need to persuade Finn. At last, without hesitation, she picks up the phone.

"Hello, Finn? Alessia here," she announces.

"Alessia, how nice of you to call. What can I do for you today?" Finn answers with an over-the-top welcoming tone. While speaking, he quickly signals his admin into his office.

"I've an Eau Suprême task. I was hoping you could help me with it," she responds.

"Well, my diary's open the next few days. I'll be glad to help in whatever way I can." In more convoluted sign language, Finn signals his admin to get Karl on the phone.

"Do you know about Hooksiel?"

"Scheisse!" Finn yells into the phone. "Don't tell me you need me for that abomination."

"Actually, yes."

"Alessia, that problem has gone on for ten years. No one's been able to fix their water. Everyone who gets involved ends up committing political suicide."

"Finn, now hear me out," she demands. "The problem with Hooksiel has always been the cisterns. The EU convinced the town to install new water pipes in all one hundred and seventeen homes. They need a water supply to test the lines. If you solve Hooksiel, your reputation will skyrocket."

"Okay, okay. So what does this involve?"

"I'm planning to install a small desalination system on the coast. It will deliver fresh water for the testing and score me a few political points, particularly in Spain."

"And if this goes down the sewer?" Finn argues. "Where does that leave me? My position here is by appointment. My superiors can remove me at any time."

"In the worst case, we can blame either the new pipes or the desalination system. In the best case, you'll look like an upcoming problem solver. Either way, I'll be glad to execute an EU assignment letter, so you can toss the blame upstairs."

"Where will you acquire a small desalination system? These systems cost millions of euros and take years to build."

"Are you familiar with Elizabeth Williams?"

"No," he lies. "Never heard of her."

"Well, she and her friend have developed a new desalination technology. It's being tested in the states. I think we can build the unit for less than one hundred thousand euros. We can manufacture it and put it into production in less than two weeks."

"Will she and her friend be traveling overseas?" Finn inquires.

"Yes, they must take part. They'll report directly to you and be on site every day until it succeeds or fails."

Unexpectedly, Finn's voice changes to enthusiasm. "Well, I'll do it. Perhaps my reputation could use some political points too."

"Finn, thank you. I know there's a risk, but the risk's low and the upside's well worth the time investment. If everything goes well in the states, we'll start the project next week. I've already lined up a suitable fabrication company. I'll email the file and authorization letter to you today."

Finn punches line two. "Karl, who do we use in northern Germany for thorny tasks? I've got a target in Hooksiel. I want them ready for next week."

It's late afternoon and Bob's worked straight through the day. He's feeling the stress of his new responsibility. Putting the final touches on his email report to Matt, he writes:

a. *Issues: five leaks mended, three water taps replaced, control panel pump rewired.*

b. *Status: saltwater, average 34,987 ppms; fresh water, average 659 ppms; brine range from about 38,000 to 45,000 ppms in the brine tanks. Five separate tests.*

c. *Recommendations: Need to fine tune voltage and water temperature, saltwater tank cleaning process requires documentation, water tap samples require benchmarking.*

d. *Summary: Expect full production this week barring any fabrication failures.*

"Bradley, here's my report for Matt," Bob says. "What do you think?"
Bradley reads it through twice. "What'll be Matt's response?"

"He'll complain it's not good enough," he smiles, "always his first response." Bob hits the enter key, and the email flies to the AMC server.

Within two minutes Matt's entering the FAB. He looks at Bob. "We need to get the fresh water below five hundred ppms. We can interest more water companies with a lower ppm. We also need to get the brine over fifty thousand. Keep working those variables. Good job."

As Matt walks out of the room, Bob looks at Bradley. "Can I make those numbers?" he asks.

"Achieving either will have a positive effect on the other, but fifty thousand ppms is a stretch." As they consider which variables to tweak first, Conley joins them.

"Bradley, can we get dinner together tonight? I'd like you to find us a good place to eat." Turning to Bob, "I hope you don't mind, but I'd like to spend some alone time with Bradley. I'm flying out tomorrow."

"No worries." Bob winks at Bradley. "Tonight, I'll be busy studying how to be a miracle worker."

"This shrimp bisque's excellent," Conley says as he spoons away his first course. "We can't get seafood bisque like this in Ohio."

"I thought you'd like this place. Beth and I ate here the day we visited the Zoo, loved the seafood. If you have time tomorrow, the Zoo's a place you ought to visit." Bradley looks around for the waitress to refill his lemonade.

"John told me today the brothers are very pleased with your desalination design. The BRAD, what a great name. It's so nice that Bob would conceive that acronym," Conley says. "They spent less than seventy-five thousand dollars on the fabrication. They're toying with licensing more units and getting into the fresh water business, but that's a few years out."

"So it sounds like they'll take a license," Bradley replies.

"They wrote me a check for one hundred thousand dollars today. Next year the price goes up to one hundred fifty thousand."

"We gave them a discount?"

"We agreed on a special first-year price. We acquire the CAD files, fabrication design, and they pay all your expenses."

"So Dad, what will you charge the next customer?"

"Based on the research Mark Conti and I did, we consider the BRAD a smaller installation, more useful for local needs. Until we can expand to meet city and industry capacity, one hundred fifty thousand dollars per year is a fair price." Conley pushes his empty bowl aside. "What should we do with the money?"

"Don't we have to incorporate?"

"Well, yes. I've an old friend who's a business attorney in Columbus. I'm having a meeting with him later this week, but the cost of incorporation's nominal."

"We should pay tithes on the income. This whole project is a blessing from God. We can sit on the rest. At this point my next step's a mystery."

"And what does that mean?" Conley questions.

"Beth, her father, and I are on a special assignment. We'll be going to Europe next week to set up another BRAD system."

"Who got you into this?" Conley says angrily. "Mark knows nothing about this or he would have told me."

"Dad, I can't say much," Bradley replies and puts down his spoon. He sees fear and anger in his dad's face.

"Dad, when Beth and I first met, we realized God put us together for a special purpose. The BRAD is more than an innovative business opportunity, more than college tuition. We believe we're doing God's will, and he'll take care of us. I want you to have the same confidence."

Glaring at Bradley he replies, "I promised your mother I wouldn't let anything happen to you. I'd send you to a good college, love the wife you choose, and spoil all your children. I know God's ways are best, but I never imagined how difficult it would be to release you into his hands. It challenges my faith and sometimes, breaks my heart." He says as his eyes glisten.

The waitress comes with the main course: two seafood platters, mixed vegetables, biscuits, and lots of cocktail sauce. "Both plates are the same," she says, setting the plates in front of them. "Like father, like son."

"Dad, as you know, I made a lot of promises to Mom too. Most of them are just plain common sense, but a few of them were contingent on a specific future. I'm sure neither of our crystal balls were working." Conley eats and nods, listening to every word.

"If we keep our promises to the letter, they become like vows. We don't want to be adhering to a vow. It's a bad move spiritually."

A big smile fills Conley's face, and he leans back in his chair. "I think this is the first time I've gotten a spiritual lecture from my son. It's great the man you're becoming. Naturally, I think you're right on the mark."

Taking the last bit of his scallops, he says, "Your mom would be so proud of you today. She would never have guessed you would invent something with your name, the BRAD. That shows a lot of admiration from Bob."

Conley pushes his plate aside. "Tell me about your relationship with Beth."

"We're getting along fine, both committed to our AMC assignments. Working long hours, we have little time to ourselves. Anyway, since her father's return, Beth enjoys being with her dad. She seems to be less excitable, more content." Bradley laughs. "She only takes her acid tablets when things go very bad."

Conley looks around to see who's sitting close by and then speaks in low volume. "What about your physical relationship?"

Conley sees the surprise in his eyes. "I'm just looking out for you," he explains.

"Well, we keep our affection limited. We don't talk about our physical desires, but we've agreed to stay pure in body. We don't often pray together either, since that bond seems to intensify the physical longing. Marriage may not be God's final intention, so we want to be careful."

"I pray for you both often. I know you love each other. Working together, but not getting time to yourselves must be a challenge. I was young once too," Conley laughs.

He motions to the waitress. "What desserts do you serve?"

As she recites the dessert menu, Conley stops her at blueberry pie. "Can I get vanilla ice cream with that?" She nods.

"Do you have key lime pie?" Bradley asks.

"Sorry, no," she responds.

"Then I'll take what he ordered."

"I'm so not surprised," she says with a grin and heads towards the kitchen.

Chapter 18

ON FRIDAY MORNING FRANKLIN, Bradley, and Beth pack their belongings between two "lessons learned" company meetings. The BRAD's performing well, and Bob's confident taking sole responsibility. After lunch, they say goodbye to their AMC colleagues. Matt's out of town, so John sees them off.

"Now remember," John declares, "we're having our big publicity event Monday morning. Wherever you are, don't forget to look for us on the net. We'll announce our new BRAD system as well as the contract Matt's closing today."

"Bradley," Bob says, "Thanks for your help. We achieved five hundred ppms. Matt told me it'll guarantee a fresh water contract with a Mexican water bottling company." Bob raises his fist and shouts, "We're going international!"

He then puts his arm around Bradley's shoulder with a little squeeze. "Now you will answer my calls twenty-four by seven, right? As soon as you step into that cab, you're my only support call center."

Bradley winks at John. "John, did we write a support contract?"

John fakes alarm. "Why . . . no. But Bob can take care of everything."

Bradley responds seriously. "Yes, you can call day or night. Call me when you get the brine at fifty thousand ppms."

John gives Beth a gentle hug. "Are we invited to the wedding? I'll bet you two are running off to get married."

Beth smiles. "Not this week, but I couldn't find a sweeter man."

A taxi pulls in and the cabby loads the luggage. As the cab pulls away, the driver turns to Franklin and says, "Frank, where are you off to next?"

"To the San Diego safe house. We'll be flying out of LAX Sunday evening."

"Is that lovely girl in the back your daughter?"

"She is. And that nice-looking guy sitting next to her may become the father of my grandchildren."

As everyone talks about their stay at AMC, the undercover cab driver maneuvers through the back streets of San Diego. After thirty minutes he stops at an alley garage door. The door opens when he honks, and he steers into a deep tunnel protected by armed security guards.

"Finn, it's Alessia," the caller states. "I know today's Sunday, but please call me. The desalination team will be here tomorrow morning. Let me know how soon you can arrive."

Finn sits at his desk, screening her call. He and Karl are enjoying German beer and plotting how to turn this desalination task to their advantage.

"I'll personally drive Williams and her boyfriend to Hooksiel," Finn states. "I think I can make small talk for five hours. Is the hostel prepared? We'll only get one night away from the hotel."

"Yes," Karl answers. "It's a private room hostel. I've had it wired for audio, video, and installed a sensor to expose communication devices. There's a ground level bedroom window for easy access. I've hired two experienced men."

"When's the event?"

"I directed the men to wait until almost dawn. We may get something valuable on audio or video during their stay. Oh, and I told them the bodies must disappear."

"Good idea, but either way, those two are dead!" Finn flies into a rage. "I want them dead and out of my way." He jumps up and pushes things off his desk, slams his chair back against the wall, and paces around the room cursing.

"When I say dead, I mean dead! No failures." Karl sits uneasily waiting for his tirade to subside.

Stomping out of his office Finn yells at Karl, "Clean up that mess and get me plane tickets to Brussels for tomorrow morning."

Karl, disgusted, snakes around grabbing papers and debris from the floor and tosses them back on the desk. He phones the Swiss office's concierge to arrange Finn's travel schedule. Karl's relationship with Finn has deteriorated over the last few years, but he still puts up with him. Nevertheless, Karl has reached the point of hoping to add Finn to his target list.

The British Airways flight taxies into the BRU main terminal. Bradley and Beth slept through the trip. Franklin spent much of the night on the phone with his Australian colleagues, Vince, and Alessia. He's nervous about this trip. He prefers not to spend time with Alessia and open old wounds. She slipped a picture of them in the Peace Corps into his pocket at the Zoo, but he's not sure it's wise to renew their relationship.

"You guys finally awake," he asks the sleepy heads.

Bradley and Beth nod, but the look in their eyes isn't convincing.

"G'day mates," Franklin teases with an Australian drawl, "Alessia will meet us in Concourse C. We'll look for an EU office at gate thirty-seven, after we collect our luggage."

"Dad, please, no Australian lingo," Beth yawns. "What's the time here?"

"It's 7:45 a.m. local time. Let's look sharp and be careful."

Lina Martens meets them as they exit the jetway. "Have a good flight?" she asks. "I've already arranged someone to collect your luggage, so follow me and I'll drop you at the EU airport office."

As Lina opens the airport office door, Alessia cordially greets Beth and Bradley with handshakes. Then she puts her arms around Franklin and kisses him on both cheeks. Her affection towards him makes him uneasy. While making small talk, Finn comes smiling through the door.

"It's been a while since I've been to this office," he says cheerfully. Walking directly to Beth, he says, "I'm delighted to meet you, Ms. Williams. This must be your young friend, Bradley. And you," he says, eyeing Alessia and Franklin, "you must be Franklin Williams. Alessia speaks highly of you."

Alessia introduces, "Please meet Minister Finn Schweitzer. He's the Swiss Government Minister of Environment, Transport, Energy, and Communications."

"Please, just call me Finn."

"Now Finn and I have made all the arrangements," Alessia announces. "You'll be staying at the Atlantic Hotel Wilhelmshaven about fifteen kilometers from Hooksiel, Germany. Hooksiel's near the North Sea and beautiful during summer. It's about a four to five-hour trip from here, but Finn will drive you. Incidentally, he only rents the most comfortable autos. On the way he can give you the background on the challenge of Hooksiel."

"And a big one it is," Finn adds, winking at Beth.

"Franklin stays with me. We need to catch up on old times, but will join you when the fabrication's finished," Alessia says with confidence. Putting her arm on Franklin's shoulder, she adds, "I need to talk with Finn a moment alone. Would you three mind waiting for your luggage in the corridor?"

After the door closes, Alessia turns her back to the glass entrance and faces Finn. "Finn, we haven't worked together very often, but I know you're a smart and perceptive person. If you don't know already, there's some shady history between Frank and me. Please let nothing slip when you're with Beth and Bradley. I don't want them to suspect anything and I don't need a scandal. Promise me."

Finn does not reply, so Alessia insists, "Well, promise!"

"Ok, Alessia," Finn says with a serious look, "I promise."

Finn assists loading luggage in the rented E-Class Coupe Mercedes. As Beth pulls her suitcase up to the car, the latches spring open, causing the case to come ajar. She quickly pushes her clothes back inside while secretly activating a signal device. Finn grabs her suitcase with a simple remark, "Time to replace this one."

Touring through Belgium into the Netherlands, Finn talks non-stop about the landscape, history, and folklore. He feigns being the perfect host.

At a road-side tavern near Groningen, the Netherlands, Finn pulls off the road and parks. "I have to make a phone call," he says. "We'll take lunch here. You're welcome to find a table. I'll be in soon."

As Beth and Bradley enter the tavern, Finn pulls out his mobile. "Karl, check the history on Franklin and Alessia. There's something dark to discover. Elizabeth may be Alessia's daughter. I want to know everything about their relationship. Ready for tonight?"

"Tonight's set. I've already checked into Franklin's background and couldn't find anything," Karl replies.

"Well, check again!" Finn insists angrily. He hangs up the phone and dials a different number.

"This is Karl," Finn lies.

"Oh, yes. How can I help?" a female voice responds.

"If Alessia asks, the Atlantic Hotel made a mistake on Ms. Williams' arrival. They'll be staying at a hostel tonight and moving to the hotel tomorrow. I've taken care of the change already. No reason to mention this."

"I understand," Lina answers. She hangs up the phone and promptly alerts Alessia.

After lunch Finn brings Beth and Bradley up-to-date on Hooksiel. Going through the history from the 1900s, he finally gets to the city's water problem.

"The older section of the city has always gotten their water from basement cisterns. About forty years ago, when city development was booming, the city tapped into the Hooksieler Binnentief. It's the fresh-water lake that borders the city. It's fed by the Neues Inhauser Teif, a long river. The water's clean and good tasting.

"The older home owners refused to connect to the lake water because the city levied a monthly tax. However, over time the cistern water has deteriorated. The water quality is technically safe, but the odor and color's offensive.

"The home owners asked the city to correct the water. They believe local industry affected the cisterns, but no one has found any problem. After they filed multiple lawsuits, the city went to the German Bundestag seeking resolution. The government forwarded the lawsuits to the EU. It's been ten years, but no one has found a solution.

"To help solve the lawsuits, last year Alessia convinced the city to install new water lines at no charge. By removing the original cistern piping, the new lines can be connected to a new water source. The city agreed to let Alessia connect a test desalination system at the EU's expense. If the desalination plant's successful, the city may also use the water and lower the tax."

"So just the homes with cisterns have bad water?" Bradley questions.

"Apparently so," Finn answers, "although there were about six households who claim to have no problem with their cisterns."

"I assume the homes with bad water drink bottled water," Beth suggests. "It must be the water for the kitchen and baths that we're replacing."

"You're right," Finn replies. "This isn't about drinking water."

For the next two hours, Finn continues to describe the countryside. Beth and Bradley were already suffering from jet lag and now, from Finn's incessant narration. They say little and are thankful when Finn stops in front of a building in the older Hooksiel section.

"Are we inspecting a cistern?" Bradley asks.

"Unfortunately, a mix up happened at your hotel. Please accept my apologies. I've already contacted Alessia's office. I found another place for you tonight. It's a typical German private hostel, a little small, but neat and clean. Tomorrow, the Atlantic Hotel will honor your reservations. Call the hotel in the morning and they'll collect you. Let me first go inside to be sure it's proper."

Finn enters the building. Beth and Bradley look at the street's housing. The structures are old and dilapidated. The brick street is full of holes.

"Not too safe," Beth whispers.

"We're well taken care of," Bradley whispers back, pointing up to the heavens.

Finn returns with a pleasant smile. "It's spotless and cozy. It should be fine for one night. Let me help you with your luggage."

Finn sets the luggage out on the cracked sidewalk. Beth pulls up the handle and her luggage pops open again. She reaches over and snaps it shut while disengaging the signal.

Finn looks over and smiles. "Again, I apologize for the mix up." He closes the trunk lid and hops into the car. "I have a few morning meetings in Berlin before I fly back to Bern, but you're welcome to call me anytime." As he drives away his face lights up with delight.

The next morning Finn arrives at the Berlin Deutsche Bank for a meeting with fifteen minutes to spare. He calls Karl for an update.

"So tell me everything went as planned," he says.

"It's still early," Karl responds. "You wanted a disappearing act. That takes extra time. I'll review the video and call you when they call me." Karl says as he hangs up. He's still trying to track down Franklin's wife and settle the question about Elizabeth's birth mother.

After the bank meeting, Finn catches a cab to the Berlin Apollinaris offices. Apollinaris is a naturally sparkling water, bottled in Germany, and owned by Coca-Cola. He's meeting a marketing vice president who's supported his agenda of protecting Europe's water industry. He's hoping to meet other company executives to widen his influence.

As the cab travels across town, Finn's phone rings. He doesn't recognize the phone number, so he ignores the call. In less than a minute, the same number rings again. He ignores the call again, but studies the number. It's an overseas 011 international code. He can't remember giving anyone from across the pond his private number.

Exiting the cab his phone rings again. He reluctantly takes the call.

"Hello, this is Finn."

"Good morning, Finn," Beth says with a cheery voice. "We switched to our hotel and will meet with the fabrication team at 10:30. We hope to

have a new CAD design by the end of today. I'll give you an update later."
Beth promptly hangs up.

Finn stands on the sidewalk motionless. His first reaction is disbelief, followed by anger. As he shoves his way into the Apollinaris building, he kicks over a trash receptacle inside the entrance. Two security guards rush over, but he waves them back and pulls out his government credentials. Breathing heavily, he sits in a chair and grabs his phone.

"Karl, have you heard anything yet," he says.

"No," Karl responds. "Are you all right?" Karl can hear the stress in Finn's voice.

"I wish I was with you right now," Finn growls. "I'd make it slow and painful."

"Finn," Karl raises his voice. "What's with you?"

"I just hung up with Williams. She called to let me know everything's on schedule."

Karl's shaken. He quietly waits for Finn's fury.

Finn's enraged and fires questions at Karl. "Who knew where they were staying? Was anyone watching the video last night? How did they escape again?" Now yelling in the Apollinaris reception area, Finn says, "Tell me we at least got several juicy pictures for blackmail?"

"Okay, okay," Karl replies. "Calm down." Karl gets his thoughts together. "Look, no one knew their location except you, me, and the two guys I hired. Someone I trust watched the live video all night and I watched the first part of the recording this morning. We can trust the guys I hired. I've used them before."

"Someone's trying to make a fool out of me," Finn sputters. "I bet it's Alessia. Find out everything. I'll call you again later."

Stepping into an Apollinaris third-floor conference room, Finn immediately senses this group isn't friendly. There's no warm greeting from anyone and no introductions to the unfamiliar faces.

A stout executive in an expensive suit and a bright red tie speaks in German. "Minister Schweitzer, please sit here by this computer monitor. We're watching an interesting news video from California."

Finn sits next to a serious-looking engineer who pulls up the news report regarding the BRAD launch at AMC. He watches the system being described in English by a technician named Bob. At the end of the narrative, the CEO gives credit to Beth Williams and Bradley Truman for their patented invention.

Finn's bewildered, but the man with the red tie scolds, "The best part is coming. Pay attention!"

Finn looks back at the screen as the local news commentator introduces a video clip of Alessia Amato.

> "I'm Alessia Amato in Brussels, Belgium. I'm the leader of Eau Suprême, an organization in Europe dedicated to protecting our fresh water resources in the shadow of climate change. We're bringing the BRAD to Europe to resolve water issues in one of our German cities. The project is under the capable leadership of Swiss Minister Finn Schweitzer. Minister Schweitzer will report on the BRAD's success when the project's completed."

Chapter 19

BETH AND BRADLEY SIT at Beth's hotel room dinette, amazed at the AMC news release. With no hesitation, they stand and slap a big high-five. It's followed by an unstable hug and dance that circles the room.

"What a great presentation," Bradley says with excitement. "Bob did an excellent job describing the system. I'll call him right now."

"Our meeting is in about ten minutes so don't talk long," Beth says, fussing with her hair before the mirror. After Bradley's acclamations to Bob, they leave for a hotel conference room on the second floor mezzanine.

"Guten Morgan," an older gentleman stands and addresses them in broken English. "I'm Leon Fischer from Wilhelmshaven Fabricating Ltd. My team does not speak English and I'm not too good. I know a little."

"Thanks for coming," Bradley says after introducing Beth and himself. "Do you understand the software design we sent you?"

"Yes, very well," Leon replies. Pointing at one of his men, he says, "Jonas went to the Hooksiel location."

Opening up his handwritten notes, Leon describes several changes required to link the system to both the North Sea and the city of Hooksiel.

In Berlin, Finn's face is becoming an angry red, but the red tie executive demands Finn stay silent. "You can answer when we're done!" he declares.

He struts around the conference table. "I'm the division leader of Apollinaris, and until this morning I've looked the other way regarding your method of protecting Europe's resources. Your tactics are harsh and illegal, but you've managed a meaningful response while others just make noise.

"Nevertheless, all of us in this room have observed this news video. Right or wrong, we've come to several upsetting conclusions. We believe you've become more interested in your political career than protecting the water bottling industry. You've disrupted many strategic plants around the globe, but can't seem to stop two kids with a new and dangerous technology.

"We assume you're protecting this Williams girl, because her father's a former lover of Alessia. We suspect you have a pact with Alessia and have permitted this BRAD system to come to Europe."

Finn twists in his chair, negatively shaking his head to the accusations. He wants to defend himself, but each time he speaks the man insists, "That's enough!"

A ringing conference room phone causes the division leader to stop. He nods to the marketing vice president.

"Hello, you've reached Apollinaris headquarters," the marketing vice president answers.

"This is Stefano from the Solé bottling plant in Nuvolento, Italy," says a voice in broken English.

The vice president nods to the man with the red tie who says, "Put the phone on speaker."

The voice asks, "Is Finn present?"

"Yes," Finn answers, relieved to hear his friend's voice.

"Finn, this is Stefano. Today's news from California troubles your Italian friends. We don't understand what you're doing. This technology can't come to Europe. It may be a small system, but it undermines our control. We agree with Apollinaris. You should resolve these matters without delay. In the meantime, we're holding our support funds in escrow. Ciao."

As Finn sits shocked by Stefano's phone call, the man with the red tie gets nose to nose. "Tell us how you've already planned to fix this problem," he says, his face now as red as his tie.

Finn fires back, "This is about Alessia . . ."

"Shut up!" the man yells. "We don't want to hear your pointless rhetoric. Tell us what you've already devised to fix this problem."

Finn wishes Karl was in the room. Karl would have this bunch on their knees begging for their lives. Looking up at the man Finn replies harshly, "I've paid the fabrication leader to ruin the system at Hooksiel. I want to go now!"

The red tie moves back and the marketing vice president opens the conference room door. Finn jumps out of his chair, hurries to the door, and bitterly glares at the marketing vice president as he exits. He tries to calm down in the elevator, but his face remains tight and flushed. His hands are shaking as he exits on the main floor. Demanding the receptionist promptly arrange a cab, he sits in a customer chair with his face in his hands.

He's furious. After everything he's undertaken for those bigheaded bottling companies, they treat him like dirt. He and Karl will finish what they started and take revenge on those Apollinaris idiots at the same time.

Bradley and Beth escort the fabrication team to the dining room for lunch. Leon's design changes were very insightful and Bradley's extremely pleased. Since the hotel serves a large buffet, Bradley and Beth treat the team. The three German specialists jabber away while helping themselves to the unlimited seafood. Beth accompanies them and successfully communicates using lots of hand motions.

Leon pulls Bradley aside. "Show me the washroom please," Leon requests.

"Of course. There's a sign right around the corner," Bradley replies motioning to the hallway.

"Show me the washroom." Leon insists.

"Yes, okay, right this way." Bradley leads Leon to the men's room door.

Leon looks around and whispers, "Follow me."

Bradley follows Leon into the washroom. Leon checks the stalls to make sure no one's present. Then he turns on a sink faucet and steps close to Bradley.

"I like you and the girl," Leon mutters in his best English. "I would like to do this job, but it's political. They paid me five thousand euros to make your system fail. If it works, I'll lose Switzerland's business."

Bradley's quiet for a minute and then speaks softly. "Please change the design as we agreed. I need your expert help. I'll report tomorrow how to make it fail."

Leon smiles, "Is this what Americans call a 'win-win'? Yes?"

"I hope so," Bradley says. "Thanks for your honesty."

As the two walk back to the table, Beth and the German-speaking team members are all smiles. The food's good and the beer's better. After two or three trips to the buffet bar the fabrication team slows down and Leon announces they must leave.

"Bradley, we've a lot more planning to do," Beth complains.

"It's okay. Leon understands the changes to the design. He'll finish it this afternoon and we'll meet again tomorrow. We can complete the review and fabrication discussions then."

Everyone shakes hands and each team member manages a 'thank you' in broken English. Leon gives Bradley a surprising hug and asks, "So we can meet tomorrow at 10?"

"Yes, right here again. Everything will be ready," Bradley replies.

As the group walks to the exit Bradley says to Beth, "We've got a lot of work to do. Get your father on the phone. We need a Plan B."

Finn drives nonstop to his Berlin hotel suite and pours a kornbrand. The alcohol calms his nerves, but not his determination. He sits down in a comfortable chair and presses the speed dial on his phone.

"Karl, what have you found for me?" he asks as soon as Karl says hello.

"I've tracked down the hospital where Elizabeth was born," Karl announces.

"Forget that," Finn yells. "Why aren't they dead?"

"The two checked into the Atlantic after midnight. In reviewing the video footage, it seems they exchanged places with two imposters. I can't locate the men we hired, so I expect someone captured them."

"Will they talk?"

"They know nothing, so I'm not worried. I paid them to terminate two renters, but they don't know why. I never contacted them directly," Karl states firmly.

"How did anyone know where they were?"

"No idea," Karl sighs. "I positioned a GPS sensor at the door, but it detected nothing. We've listened to their conversation and can find no clues."

"Do we have any compromising video? A few naked shenanigans or anything we can use for blackmail?" Finn questions.

"No. According to Fredrick, they were more passionate at that landfill in the states. Last night they studied a map of Hooksiel reviewing the coastline site options for their system. They baked the pizza you supplied and around 21:00 they put on commonplace night clothes, nothing to suggest homemade entertainment. Elizabeth crawled into the bed and he went to the couch. While he was reading, Elizabeth came to see him. For the next hour they talked about the Bible and prayed together. Then both returned to their own beds with lights out.

"Around 23:30 he got up. He woke her and asked her to come to the basement to see if there was a cistern. He only turned on the basement lights. They were in the basement for about fifteen minutes. They came back with lights out, but the video shows the shadows of two people. One went to the couch and the other to the bedroom. I think that's when the exchange happened."

"I suppose there's no video in the basement?" Finn inquires with a nasty tone.

"No," Karl responds without further comment.

Finn becomes angry, "Why didn't you plant surveillance in the basement? Again you mess up and they slip away." Karl clinches his fist, but does not reply.

"So did you firm up the deal with Leon?" Finn asks.

"Yes. I met with him myself. He knows he must make a serious error, or he will lose his business in Switzerland."

"It better be a permanent error. I don't want any fix to be possible."

"I agree."

"Leon's our backup plan, but I need you to get three things done yesterday," Finn says impatiently. "I want you to eliminate Elizabeth and that guy, Bradley; destroy that California brine system; and demolish whatever fabrication's completed for Hooksiel."

Finn pauses and then yells, "Karl, if you slip up again, you're finished with me. I'll make sure you're someone's target. You understand!" Karl abruptly hangs up the phone, so Finn makes another call.

"Hello, this is Karl," he lies.

"Give me just a minute," Lina responds. After a few seconds, she says, "It's okay."

"I need the hotel room numbers for Elizabeth and her friend."

"Send as usual?" she asks.

"Yes and give them a reason to leave the hotel for a while."

"I think I can do that."

"Have you heard any rumors lately?" Finn asks worried about himself.

"Our office received a call after lunch from someone at Apollinaris. Alessia was very upset."

"What did they talk about?"

"I don't know, but after the call, she called Interpol."

Finn drops the mobile in his lap. Grabbing it, he turns it off. He decides if he's going to get a job done right, he'll do it himself. He dials an old friend in Berlin.

"Franklin, based on that Apollinaris call, we must talk with President Diego right away," Alessia urges. "Let me set up a meeting this afternoon."

"I thought we needed time to make our meeting look like his idea. If we take initiative, will we get the same results?" Franklin says with a worried tone.

"Let's hope our game takes him off guard. Maybe it'll make him more receptive," Alessia responds as she dials Diego's office.

"Diego, do you have time to see me today?" she asks.

"Whatever!" he barks. "I've a meeting scheduled in thirty minutes, so come now."

Stepping into his office, Diego sees a stranger with her and says, "Let's keep this short."

Alessia introduces Franklin to Diego. The president steps back thinking out loud, "Williams, Williams? Wasn't there a Williams who survived that awful shooting at a nearby hotel a few months ago? Any relation?"

"Yes, she's my daughter," Franklin states.

"I'm so sorry," Diego relaxes his prickly disposition a bit. "There's been several rumors about violence associated with this climate change business. Did they arrest the people who committed those murders?"

Franklin shakes his head, "Not that I'm aware."

Alessia takes Franklin by the arm. "Diego, Frank and I are renewing an old romance. I'm considering moving to the states with him and his daughter. If it happens, how much advance notice would help you?"

"Well, we'd need to replace you right away. I think six weeks is suitable. We'd certainly miss you, but if you're happy together, that's what's important. When are you planning to leave?" Diego gives Alessia a wink and a relaxed smile.

"No date yet, just considering all the options," she responds. Then with an urgent appeal, "If I resign, please don't hire Finn Schweitzer to replace me. We suspect he may be behind much of the violence."

"You don't get to vote on your replacement," Diego remarks. "I know Finn has a checkered past, but those days are gone. He'd be one of my first choices. What makes you suspicious?"

Alessia nods to Franklin. "Mr. Gonzalez, I recently worked on a desalination project near San Diego, California. While I was there a fellow chemist faxed classified information to an international number. They traced the fax number to Finn's office in Bern. Two days later the chemist received a FedEx package with enough explosives to destroy the building."

"I hope this isn't anybody I know," Diego responds with a worried look.

"Tobias Gruber."

Diego clutches his leather office chair, sits down, and wipes his brow. "I recommended Gruber for the AMC position. Well, Finn chose Gruber for my recommendation. Where's Gruber now?"

"He's in the custody of the FBI in San Diego."

"Can I confirm your story with someone?"

"Sure. Call the San Diego FBI office," Franklin replies, "but I don't know which agent you should contact."

Diego walks to his office door and calls to his admin, "Get me the FBI in San Diego on the phone. I don't care what time zone California is in."

Walking back to his desk, he snaps, "I want to know the truth. I recommended Gruber."

When his phone buzzes he connects to the San Diego office. He takes notes as he listens to the agent relay information. Except for writing on his pad, he's expressionless. After several minutes, he thanks the agent, and hangs up with a deep groan.

"You're right. Gruber's a criminal. They're suspicious Schweitzer's involved with him and another man named Hans Guttmann." Diego stands at his desk. "Alessia, I hope you're not leaving soon, I'll need your expertise for this."

"Like I said, no plans at present," she smiles. "Let me know what you need."

Alessia and Franklin walk back to her office. "I think renewing our romance was fun. It helped to break the ice with Diego. He acted like he knew nothing."

"I agree. He seemed genuinely outraged," Franklin replies. "I didn't want to come to Brussels, but I've enjoyed every minute on this masquerade with you."

"So you're thinking of attaching the nickel plating to the outside," Franklin says sitting in Lina's office, "It'll work, but not on steel."

"So what materials will work?" Bradley questions.

"For what you are suggesting a synthetic plastic polymer's the best choice, think PVC pipe. Now, depending on the PVC thickness it may not winter well. Leon should be able to recommend."

"Do they use PVC in Germany?"

"Oh, sure. The chemical composition's probably somewhat different, but you'll get the same results. You must adjust a few of your settings, since a polymer surface will change the water flow."

Lina interrupts Franklin pointing at his mobile. "Listen, Lina's here and wants to talk, so I'll put her on." Franklin hands his cell to her.

"Glad I caught you. You need to move out of your hotel rooms. In fact, move to a different floor. Leave a few personal items, so the room doesn't look vacant. The heat's turned up and you're both targets. Please be careful." She ends the call.

"So will this Plan B work?" Beth asks. She's been listening to Bradley's side of the call.

"Your dad thinks so," Bradley responds, "but right now we have to move. We need different hotel rooms. Lina thinks we're in danger again tonight."

Beth hops around, grabbing her things. She reaches for her pills and pops one in her mouth. "Just in case," she says to Bradley with a smirk.

"Sit tight," he says. "I'll be right back."

Bradley takes the back stairs down to the hotel kitchen. Searching around the kitchen floor, he finds what he anticipated. He marches out to the reservation desk. "May I speak to the manager?" he asks with a serious look.

A well-dressed woman appears from an inner office and says pleasantly, "Mr. Truman, how may I help you?"

Taking the manager aside he speaks quietly, "My friend and I have large bugs in our rooms."

"That's impossible," the manager challenges. "Those rooms receive regular treatments. I've had no other complaints."

"When was the last time our rooms were treated?" Bradley inquires.

"Well, I believe it was about six weeks ago."

"I think you should do it again," he says. Opening his hand, he holds out a large insect, the oversized cockroach he found in the kitchen.

The manager steps back, looks around, and inconspicuously closes his hand with hers. "I'll find you two rooms on another floor for tonight and we'll spray your rooms. No need to remove all your things. And no need to cause alarm to our other guests. I'll not charge you for tonight."

Bradley nods with a smile. "Shouldn't you have cameras in the rooms to find insects?"

"We have hidden video in the hallways only."

"If you don't mind, I'd like to see the hallway recording in the morning. Just to be sure someone sprays," he says seriously.

"Why, of course, Mr. Truman," she responds. "Let me get your new room keys, so you can move your belongings."

Within an hour, Beth and Bradley finish moving their essentials to new rooms on the second floor. As they sit in Beth's new room praying for their safety, Lina calls.

"Have you moved your rooms?" she questions. Beth confirms. "Great, you'll be fine. I've booked a table for you at a historic restaurant in Wilhelmshaven. The German food's supposed to be exceptional and they'll have live music tonight. Enjoy yourselves. I don't want you stewing over anything, but don't return before 21:00."

Bradley arranges for a taxi. For fun the two dress in silly outfits. Beth wears a sleek black dress covered with sequins, arrayed with a bright red silk scarf. Bradley waits for her in the lobby wearing his traveling outfit, a double breasted gray pin stripped suit, accented by a newly purchased navy blue bow tie.

"We look like two people in a B movie," Beth laughs as their taxi arrives.

Sitting in a coffee shop across from the Atlantic Hotel, Finn watches the young couple strut to the taxi and drive off. He finishes his coffee and carefully picks up a large paper bag. Even with his frantic four-and-a-half-hour drive from Berlin, he's contented with himself and his plan. He wishes he'd taken personal control sooner.

He enters the Atlantic Hotel Wilhelmshaven and marches to the elevator as if he were a guest. Getting off at the second floor, he searches the hallway for a chambermaid's closet. Finding one at the end of the hall, he opens the door and searches through the closet for a master key. Finding none, he takes the stairs to the third floor and follows the same routine. At last, on the fourth floor a maid's left her master key in an apron pocket.

Returning to the third floor, he pads back and forth through the hallway a few times unaware of the hidden cameras. He listens for noises suggesting someone may step out of their room. With no activity, he uses the master to enter Beth's room unseen. Cautiously setting down his bag, he pulls latex gloves from his pocket and lifts a metal container from the bag. After checking the electronic timer, he slides the container under Beth's bed. With the master key he enters Bradley's room through the adjoining room door and slides a second metal container under his bed.

He folds the paper bag and puts it in his rear pants pocket. Then taking his gloves off, he shoves them into his coat's right-hand pocket. Dropping the master key in the trash bin, he cracks Bradley's door open, checks the hallway, and sees no one moving around. He takes the stairs to the lobby, exits the hotel, and crosses the street. He's all smiles; his problems are solved. Time for a quick coffee before he drives back to Berlin. He discards the paper bag and other evidence from his pockets into a street trash container. As he re-enters the coffee shop, a disguised man with a crooked nose walks into the Atlantic Hotel.

Chapter 20

AFTER A LONG DRIVE and little sleep, Finn prepares to check out of his Berlin hotel. In a few hours, he'll be back at his office in Bern. He and Karl will plot their next step. As he hands his credit card to the cashier, the TV news is blaring behind the desk. It's reporting the death of an unnamed man and woman from cyanide gas at the Atlantic Hotel Wilhelmshaven. Finn pretends not to listen. He chuckles to himself as he hops in his rental car.

Returning his vehicle at the airport rental agency, he boards a shuttle to his terminal. The cyanide gas news is playing live on the shuttle bus:

> *"The police have confirmed a man and a woman died of cyanide poisoning in the Atlantic Hotel Wilhelmshaven. Hotel video is being examined to identify the person responsible. Authorities speculate that the killer intended to poison two American guests. The Americans were away and are unharmed. They found a man hiding in a wardrobe with a loaded weapon. The woman was a hotel employee spraying for insects. Her name is not yet public. The man is Karl Muller. Muller is under suspicion for crimes in multiple countries."*

Finn is numb. He sits frozen in his seat, disoriented. In his first coherent thought he wonders if Karl's body links to him. When he drops his carryon to the floor, he realizes his hands are shaking. Reaching for it, the bus swerves into the terminal and his face slams into a support rail. Befuddled, he pulls himself up as the bus stops. When the bus doors open, he grabs his bag and runs for an airport washroom.

Rinsing his face with warm water calms his nerves. As the water soothes him he gazes in the mirror. His face is pale, his head's pounding, and a small dark lump's growing on his forehead. Unlike Karl, he never uses a disguise, but now he paws through his carryon for something to hide his face. Eventually, he gives up on his extra clothing and looks around at those in the washroom.

"Excuse me," Finn says to an athletic looking teenager. "Can I buy your football jacket and cap? You're wearing my favorite team." Finn holds out five hundred euros.

The kid says, "You bet. The jacket may be a little big, but the cap will fit fine."

Finn takes off his suit coat and shoves it into the trash receptacle. Donning the sports attire he stares at himself again. He lowers the brim of the cap a little and pulls the jacket zipper up to his chin. He hopes this outfit will let him board the plane without questions.

He speeds through the high-level security checkpoint showing his personal Swiss credentials and grabs a snack at a kiosk. Just as he reaches gate twenty-one, he hears his name mentioned. Turning to a TV monitor he spots a Swiss government representative stating:

> "After reviewing the hotel video footage, the Swiss government confirms that Minister Finn Schweitzer is responsible for the Atlantic Hotel poisoning. Schweitzer's dismissed from all duties. We will fully cooperate with the authorities for his immediate apprehension."

Finn stops. He can feel his heart thumping in his head. He turns and starts walking back towards the security checkpoint. Seeing a side door, he exits down the employee stairs and finds his way to the baggage claim. To escape he must have a vehicle. Near the taxi area, several cars are picking up inbound travelers. He waits until an idling vehicle is left alone and leaps behind the wheel. In a few minutes he's out of the airport and driving to the autobahn.

After traveling over one hundred and fifty kilometers he pulls off at an exit road. Dialing his mobile, he looks for anyone who can help him.

"Hello, this is Karl," he says anxiously.

"Who are you trying to reach?" an unfamiliar voice says.

Finn wants to hang up, but he's desperate. "I'm calling for Lina Martens."

"Who's calling?"

"Karl, a friend, she'll know." Finn's worried he's already been on the call too long.

"I'm sorry, Lina's at Interpol headquarters. I'm an Interpol officer. May I take your information?"

Finn quickly hangs up his mobile. He's distressed they'll trace the call. Tossing his phone out the car window, he wildly pulls back on the highway.

There's a knock on the door. Bradley crawls out of bed and peers through the peephole. The hotel manager is waiting for him.

Opening the door, the manager hands him two bags. "This one's yours and this one's hers. These are the remaining items left in your other rooms. You cannot return to those rooms."

"Why not," he yawns.

"Two people were killed in those rooms. Police are everywhere. You can see it on the TV news."

Bradley collects the two bags and turns on his room television. As he sits watching for the next several minutes, he catches both reports. First, he sees the one identifying Karl Muller, and then the Swiss president denouncing Finn because of video evidence.

He calls Beth's room, gives her a quick summary, and suggests, "After our meeting, let's eat lunch where we can catch the news."

At 10:00 they enter the mezzanine conference room. Leon's only a few steps behind them and comes alone.

Before he sits down, he shakes both their hands and says, "BPOL are all over the hotel. They have automatic weapons. So exciting. I had to show my papers at the door."

"Have you seen the news?"

"Yes," Leon smiles showing his nicotine-stained teeth. "They caught up with that swine Schweitzer and his bulldog Muller. I hope they lock Finn away for one hundred years." Leon stops. "That makes good news for you. We can build your design without fear."

As the three discuss certain final fabrication issues, Beth's mobile rings.

"It's Dad," she says. She steps away to talk then hurries back.

"Leon, we're getting a quick call on the conference room phone," she says moving the hotel phone closer to them. "It won't take long, and you're welcome to stay right here."

The phone rings and Beth picks up the phone. After saying hello, she puts the phone on speaker.

"This is Vince," a familiar voice says. "Who's on the line?"

"Franklin's here. I'm in Brussels."

"Bradley and I are in Wilhelmshaven with the fabrication contractor, Leon."

"Leon," Vince says, "how soon will that system be ready for testing?"

"One week from tomorrow," Leon replies. "I promise."

"That's great news. You keep your promise. Beth, Bradley, and Franklin, there's a meeting 11:00 a.m. in DC two days from now and I want you here," Vince states. "Leon, is that a problem for you?"

"No, the design is complete," Leon replies, "but they must be here for testing."

"I understand," the voice agrees. "We all want Hooksiel online as soon as possible. See you all soon." The line disconnects.

Bradley looks at Leon. "Do we have everything covered?"

"Should I make one or two systems?" he grins. "Don't you think somebody else in Europe will need good water?"

"Let's get one working first," Bradley laughs and shakes his hand. "Thanks for your honesty and patience. We'll be back next week."

Two days later Franklin and Beth exit their cab at Pennsylvania Avenue and 17th Street, NW, Washington, DC.

"I've been to DC before, but never here," Beth says to her father. "This is quite an impressive building."

"The military built the Eisenhower Executive Office Building in the late 1800s. For many years it was the world's largest office building. I've been here many times to meet with Kleaver, but not since he's occupied the vice president's ceremonial office."

After passing through the security check, they're directed to a uniformed guard who escorts them to the VP's outer office. Sitting on a large leather couch, Conley and Bradley are reviewing the incorporation requirements of BRAD Enterprises, Inc. They stand and greet Beth and Franklin as they enter.

"I didn't think I'd be seeing you again so soon," Conley says reaching a hand to Franklin.

"Perhaps this time we'll get a better chance to talk," Franklin replies. "I envy the fact you've enjoyed so many years with your son. I wish Beth and I had spent more time together."

A skinny redhead with freckles cracks the inner office door. "Five more minutes," she says.

A security guard escorting a photographer barrels through the room and right into the VP's office. Behind them an unescorted bald man, with a curly gray beard, makes his way directly into the office. The entrance door opens a third time and another security guard enters the room with a scholarly-looking woman in her fifties.

Franklin gasps, "Nancy, is it you?"

She walks over to Franklin and discreetly says, "I'm here to see Elizabeth."

Hearing that, Beth turns, studies Nancy's face, and then hugs the woman. "Mom, I can't believe it." They hold each other as tears slide down their cheeks.

Vice President Kleaver opens the door and motions them into his large office. Beth grabs a box of Kleenex sitting by the leather couch and clasps her mom's hand. The men follow the ladies. After everyone finds a place to sit, VP Kleaver gives everyone a big smile.

"I'm pleased with the family reunion," he says with a serious tone. "Now before we celebrate the great work of Franklin, Beth, and Bradley, I owe most of you my apologies.

"About twenty years ago I recruited Franklin into a scientific branch of the CIA. At the time, I didn't expect the work would become so demanding and dangerous. As a result, there were many unintended and, I'll admit, tragic consequences. I hope today, Nancy, you can accept my sincere apologies. I also hope, Beth, you'll find it in your heart to forgive me as well. Franklin, your work has always been exceptional. You and I have known each other for over thirty-five years. You're the last person I'd want to offend, so please forgive me for my misguided intentions."

For a minute no one responds. Then Beth stands and hugs the VP. "Thank you for your frankness and humility. I forgive you. I so appreciate that you arranged my reunion with Dad, and now Mom."

Conley and Bradley watch as the whole family forgives and hugs Kleaver. Bradley whispers to his dad, "There's a lot of overdue healing happening here."

"I want a picture of this moment. It's for me and no one else." The VP motions to the shutterbug. "I carry the weight of my blunders. I want a scrap book photo that reminds me to make things right if I can."

The vice president then asks Franklin, Nancy, and Beth to pose with him. During the poses, there're genuine smiles and some light-hearted jesting. After they complete the photos, the photographer's released.

The VP asks Nancy to wait in the outer office. He shakes Nancy's hand and walks her to the office door. Before she leaves he says, "Beth, I've arranged for your mom to be here a few days so you can catch up."

Turning his attention to those in his office, Kleaver says, "Now that we've completed the most important purpose for this meeting, let me update you on the foreign news. The German Police found Minister Schweitzer's dead body late yesterday near the car he stole in Berlin. From all appearances he committed suicide. In the last two days twenty-four water bottling executives from thirteen European companies resigned because of their association with Schweitzer. Alessia informs me we may not have seen the end of those dismissals. Now, about us.

"Franklin, your duties at the CIA officially end the last day of this month. Although your service is commendable, it's time I release you. Spend time with your daughter before she marries that guy in the chair."

The VP motions to the man with the curly beard who's quietly been sitting in a corner chair. "This fine gentleman is Dr. Wilson Bjorkland. Dr. Bjorkland's the deputy chairman of the US Climate Change Commission, the USCCC, as well as a senior administrator at Georgetown University. He's a good friend and has offered to assist me in making up for past sins."

Dr. Bjorkland steps to the center of the room. "Last week a member of the USCCC flew to San Diego to visit the Aquatic Mining Company lab. He came back raving about the efficiency and design of the BRAD. He believes it's a huge step forward in technology for desalination.

"Now he's recommended Georgetown buy a license and install it on the premises. The GU board agrees on the condition that Bradley Truman enroll in the university for four years. All tuition and a retainer for support services on the BRAD will be covered. How does that sound?"

Bradley smiles politely, but looks intensely at Beth. "Oh, I see," Dr. Bjorkland remarks. "Please let me go on before you decide.

"The USCCC has alerted its chairman, the honorable vice president, that Elizabeth Williams should become a paid part-time member of the commission, reporting directly to me. I'll only agree if Ms. Williams also takes a part-time teaching position at Georgetown and agrees to lecture at sister universities across the US. If you want to further your education, all tuition will be at no charge."

Big smiles appear on Bradley and Beth's faces.

"That's more than generous." Franklin exclaims.

"I'm not done yet," the administrator says. "I'll make you a similar offer, a part-time teacher and university lecturer, but no tuition benefits. With your education, I doubt you'll be wanting to take any classes, anyway." Everyone laughs.

"Thank you so much," Kleaver says. "This is not just about trying to right an old wrong, the USCCC believes Bradley has turned the desalination industry from an expensive steam-based system to an inexpensive electronic based system. I think this team's well positioned to help lead this country through the mystery of climate change."

After his remarks, the vice president and the Georgetown University administrator run off to other appointments. Dr. Bjorkland leaves the name of the registrar with them. "I recommend you contact him right away. He's expecting to hear from each of you. The fall semester is right around the corner."

Two days later Bradley and Beth sit on the Capitol Building steps overlooking the reflection pool. The Washington monument's lights are in the distance, the day's cooling after the hot sun slips below the horizon, and the coming twilight brings city lights sparkling in the pool water.

"What did you and your mom do today before she left?" Bradley asks.

"We sat in the hotel dining room and talked from breakfast until lunch. Then going to her hotel room, we took a little nap together. It was just like high school best friends. Riding to the airport I looked at pictures of her family. We've got a standing invitation to come to Oregon."

"That's wonderful. I remember the first day we met," Bradley reminisces. "It's hard to believe we've lived so much life in only eight months. In eight months you've gotten your family back and I've invented an unbelievable product. God's so good in every way and I don't deserve it."

"I know what you mean. I don't even want to think about how many times we survived death," Beth replies, taking his hand. "God's my strength, and you're my support. I don't think I can ever get along without you."

Reaching into his pocket he hides a small box in his hand. "I'm glad to hear you say that," Bradley smiles. Moving down a step he kneels before her. He opens the box and asks, "Will you marry me?"

Beth gazes at the shimmering ring and a tear glistens on her cheek. "Yes, my wonderful friend. I've loved you from the day you blushed at everything I said."

Bradley moves close for a kiss on the lips, but Beth interrupts, "Have you prayed about this?"

Bradley smiles and says, "Every day since our first date." He leans in for a kiss, but Beth stops him again and murmurs, "Yeah, but what about college and . . ."

Bradley touches his index finger to her lips. "I'm transmitting, not discussing," he says firmly. Putting his arms around her, he gives her a gentle squeeze and a loving kiss.